E.S. Gopi

Multi-Disciplinary Digital Signal Processing

A Functional Approach Using Matlab

Springer

E.S. Gopi
Department of ECE
National Institute of Technology,
Tiruchirappalli
Trichy, Tamil Nadu
India

ISBN 978-3-319-57429-5 ISBN 978-3-319-57430-1 (eBook)
DOI 10.1007/978-3-319-57430-1

Library of Congress Control Number: 2017938129

Printed on acid-free paper

This Springer imprint is published by Springer Nature
The registered company is Springer International Publishing AG
The registered company address is: Gewerbestrasse 11, 6330 Cham, Switzerland

Dedicated to my wife G. Viji, *my son*
A.G. Vasig, *and my daughter* A.G. Desna.

Acknowledgements

I would like to thank Prof. Dr. Mini Shaji Thomas (Director, NITT), Prof. M. Chidambaram (IITM, Chennai), Prof. K.M.M. Prabhu (IITM, Chennai), Prof. S. Soundararajan (Former Director, NIT, Trichy), Prof P. Palanisamy, Prof. B. Venkataramani, and Prof. S. Raghavan (NIT, Trichy) for their support. I would also like to thank those who helped directly or indirectly in bringing out this book successfully. Special thanks to my parents Mr. E. Sankara Subbu and Mrs. E.S. Meena. I thank the research scholars Ms. C. Florintina and Ms. G. Jaya Brinda for their assistance in proof reading the final manuscript.

Thanks,
E.S. Gopi

Contents

Chapter 1
Sampling and Reconstruction of Signals

1.1 Sampling of Sinusoidal Signal

Consider the signal $x(t) = sin(2\pi f_i t)$, where i=1,2,3 with $f_1 = 1$ Hz, $f_2 = 11$ Hz and $f_3 = 21$ Hz that are sampled with the identical sampling frequency $f_s = 10$ Hz to obtain the discrete sequence $x(nT_s) = sin(2\pi f_i nT_s)$ (refer Fig. 1.1). From the figure, we observe that amplitude of the discrete sequence obtained from all the three signals ends up with identical values. In fact, we get the identical sequence for infinite number of sinusoidal signals with frequencies given as $f_r = f_1 + rf_s$. The discrete sequences obtained from the signal $x_1(t) = sin(2\pi f_1 t)$ and $x_2(t) = sin(2\pi f_r t)$, with $f_r = f_1 + rf_s$, where f_s is the identical sampling frequency used to sample both the signals are obtained as follows.

$$x_1(t) = sin(2\pi f_1 * t) \quad (1.1)$$

$$\Rightarrow x_1(nTs) = sin(2\pi f_1 * nT_s) \quad (1.2)$$

Also the discrete sequence obtained from the signal $x_2(t)$ is obtained as follows.

$$x_2(t) = sin(2\pi f_r * t) \quad (1.3)$$

$$\Rightarrow x_2(nTs) = sin(2\pi(f_1 + rf_s) * nT_s) = sin(2\pi(f_1 * nT_s) \quad (1.4)$$

Thus we observed that the identical sequences are obtained by sampling the signals $x_1(t)$ and $x_2(t)$ when sampled using the identical sampling frequency. Intuitively we understand that the frequency content of the discrete sequence obtained are $f_1 + rf_s$, i.e., the spectrum is periodic with the period f_s. If the analog low pass filter

$$h_1(t) = sinc(\frac{t}{T_s}) \quad (1.5)$$

E.S. Gopi, *Multi-Disciplinary Digital Signal Processing*,
DOI 10.1007/978-3-319-57430-1_1

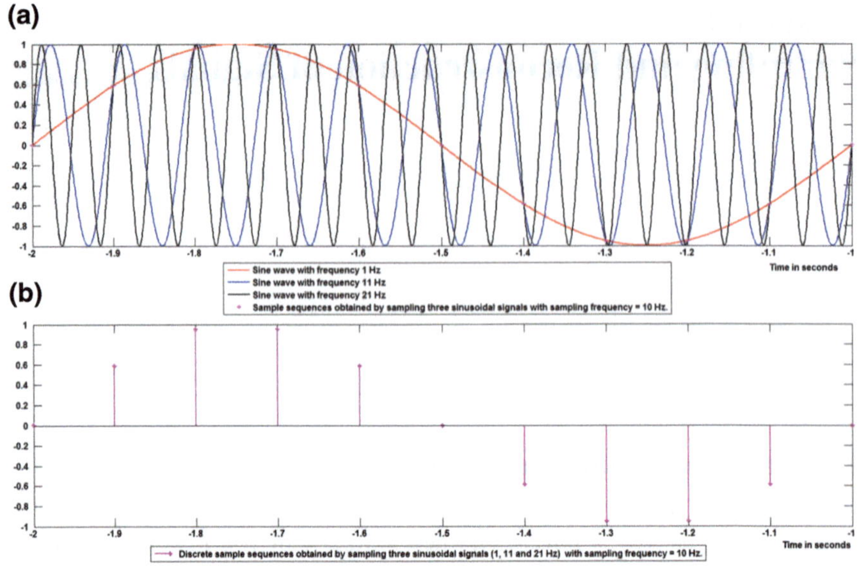

Fig. 1.1 Illustrating sampling of three sinusoidal signals (with different frequencies) with fixed sampling frequency ends up with identical sequence

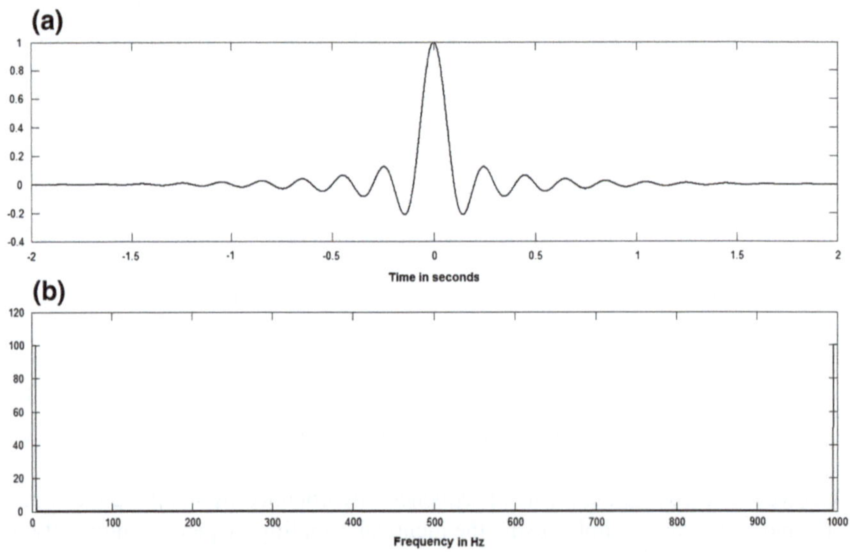

Fig. 1.2 **a** Impulse response of the Low pass filter $sinc(\frac{t}{Ts})$ (that allows 1 Hz) used to reconstruct the signal back from the discrete sequence generated in the Fig. 1.1. **b** Corresponding magnitude response

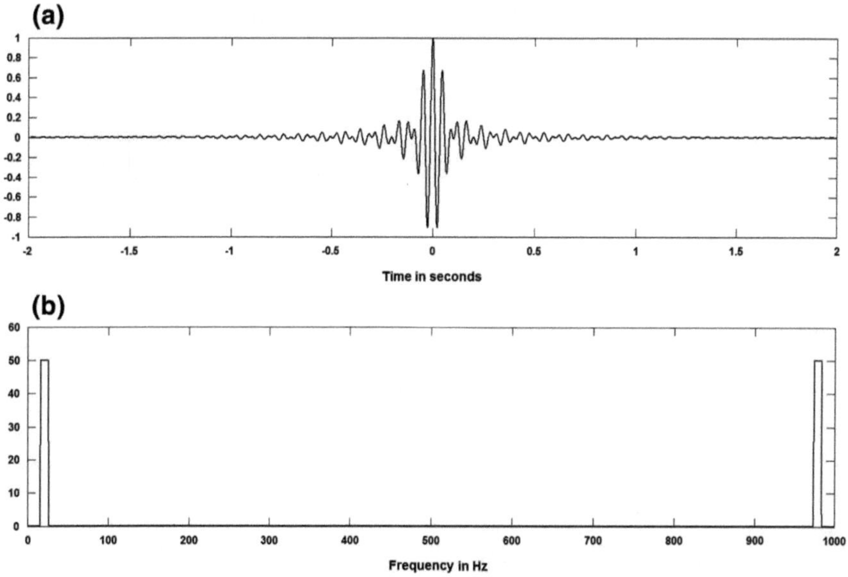

Fig. 1.3 **a** Impulse response of the Band pass filter constructed using the low pass filter used in Fig. 1.2 (that allows 11 and 9 Hz) used to reconstruct the signal back from the discrete sequence generated in the Fig. 1.1. **b** Corresponding magnitude response

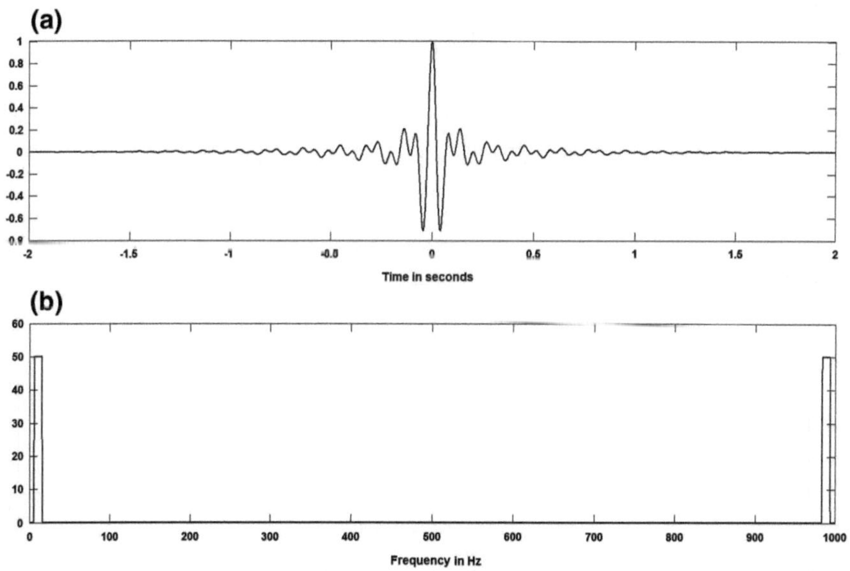

Fig. 1.4 **a** Impulse response of the Band pass filter constructed using the low pass filter used in Fig. 1.2 (that allows 19 and 21 Hz) used to reconstruct the signal back from the discrete sequence generated in the Fig. 1.1. **b** Corresponding magnitude response

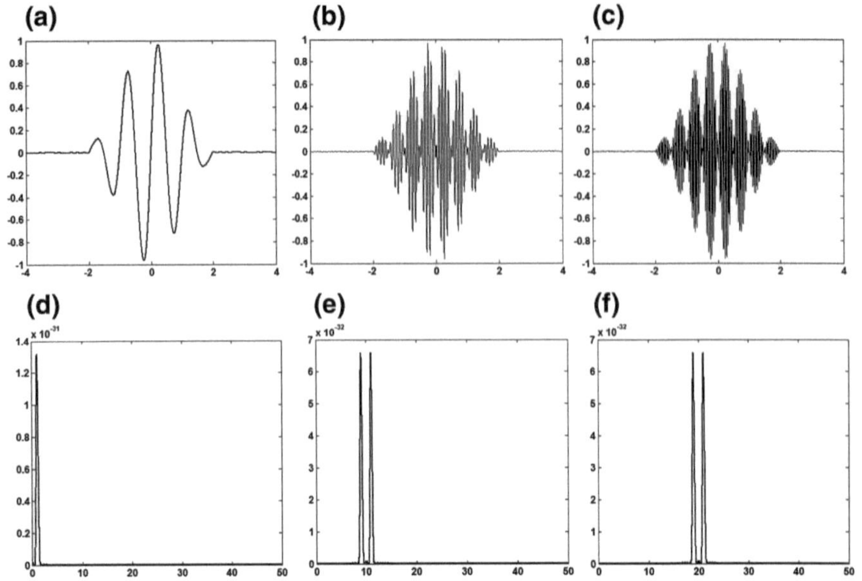

Fig. 1.5 **a**, **b** and **c** are the reconstructed signals obtained using the filter used in Figs. 1.2, 1.5 and 1.6 respectively. **d**, **e** and **f** shows the corresponding spectrum

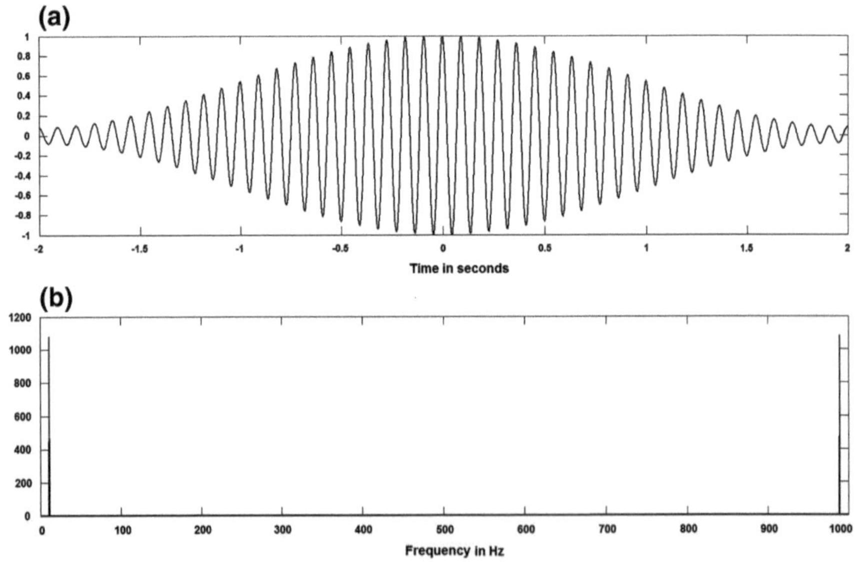

Fig. 1.6 **a** Impulse response of the narrow Band pass filter (that allows 11 Hz) used to reconstruct the signal back from the discrete sequence generated in the Fig. 1.1. **b** Corresponding magnitude response

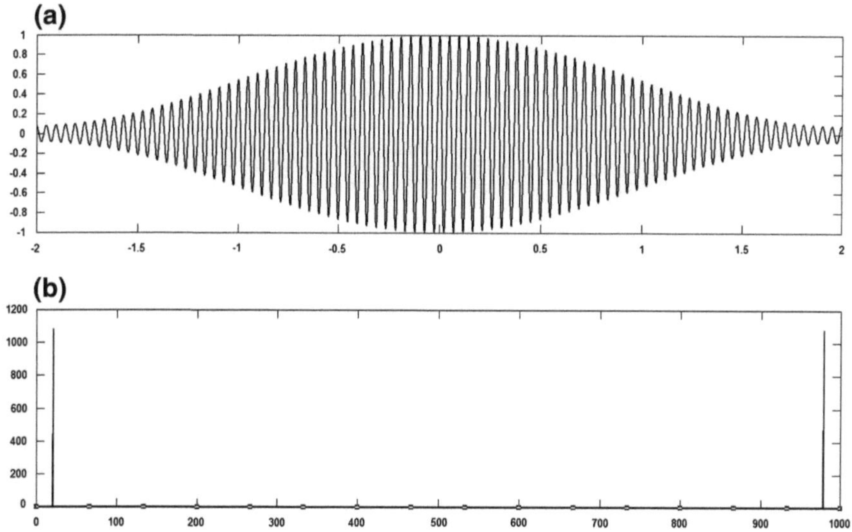

Fig. 1.7 **a** Impulse response of the narrow Band pass filter (that allows 21 Hz) used to reconstruct the signal back from the discrete sequence generated in the Fig. 1.1. **b** Corresponding magnitude response

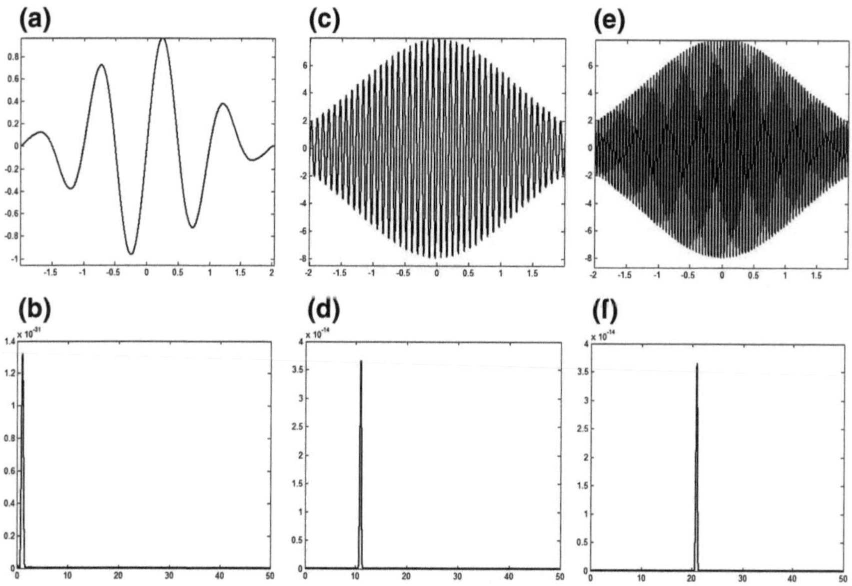

Fig. 1.8 **a**, **c** and **e** Are the continuous signals reconstructed from the samples using the filter described in Figs. 1.2, 1.6 and 1.7 respectively. **b**, **d** and **f** Shows the corresponding spectrum

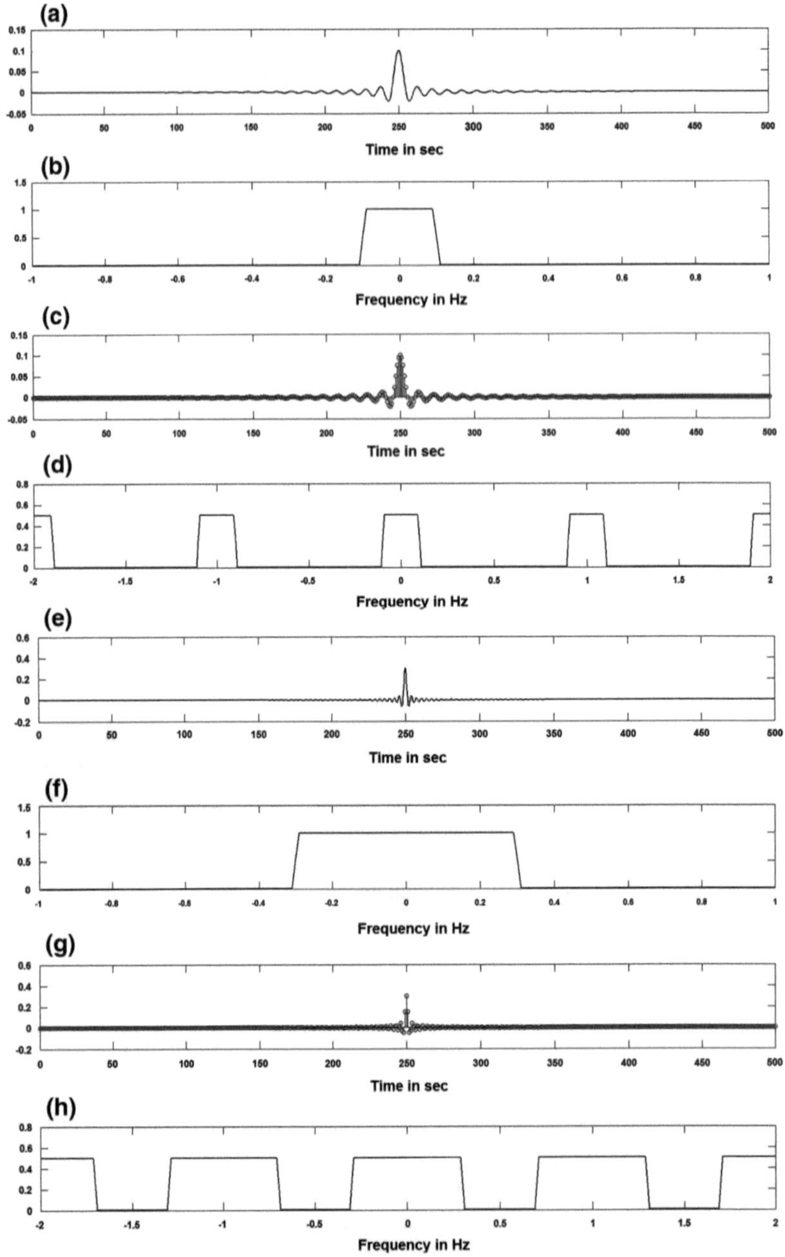

Fig. 1.9 Demonstrating nonoverlapping of spectrum in DTFT domain when $F_{max} < \frac{F_s}{2}$. **a** Low pass signal with the maximum frequency content of 0.1 Hz in time domain. **b** Corresponding spectrum. **c** Discrete version of the signal in (**a**) with $F_s = 1$ Hz. **d** Corresponding spectrum obtained using DTFT. **e** Low pass signal with the maximum frequency content of 0.3 Hz in time domain. **f** Corresponding spectrum. **g** Discrete version of the signal in (**e**) with $F_s = 1$ Hz. **h** Corresponding spectrum using DTFT

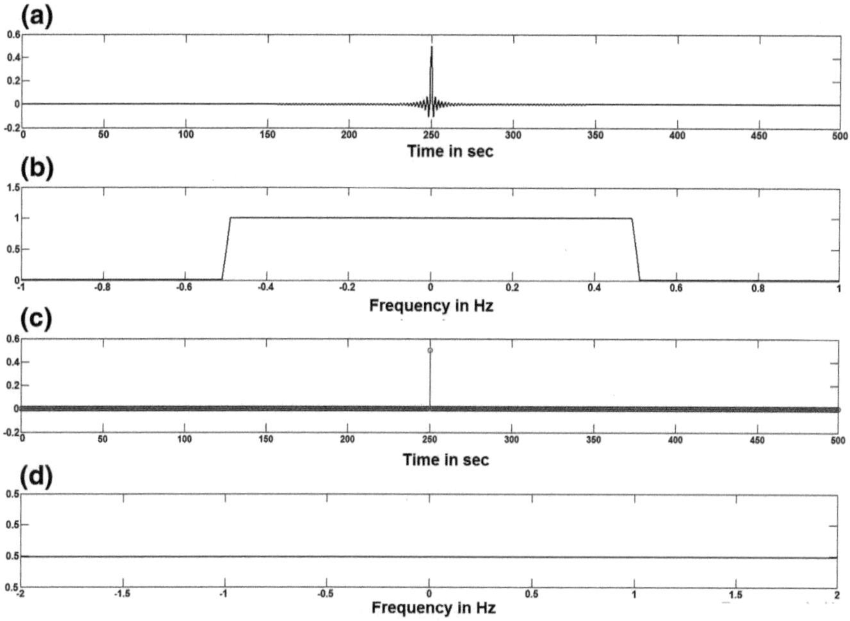

Fig. 1.10 Demonstrating marginal overlapping of spectrum in DTFT domain when $F_{max} = \frac{F_s}{2}$. **a** Low pass signal with the maximum frequency content of 0.5 Hz in time domain. **b** Corresponding spectrum. **c** Discrete version of the signal in (**a**) with $F_s = 1$ Hz. **d** Corresponding spectrum obtained using DTFT

(refer Fig. 1.2) that allow 1 Hz signal is used to reconstruct the sequence $x_1(nT_s)$, we get the sinusoidal signal with frequency 1 Hz. If the identical sequence is reconstructed using the band pass analog filter with mid-band frequency 11 Hz

$$h_2(t) = \frac{1}{2}sinc(\frac{t}{T_s})e^{-j2\pi 11t} + sinc(\frac{t}{T_s})e^{j2\pi 11t}; \tag{1.6}$$

(refer Fig. 1.3), we get two sinusoidal signal with frequencies 9 and 11 Hz. Note that the bandwidth is constructed such that it allows both 9 and 11 Hz. Similarly if the identical sequence is reconstructed using the narrow band pass analog frequency with mid-band frequency 21 Hz

$$h_2(t) = \frac{1}{2}sinc(\frac{t}{T_s})e^{-j2\pi 21t} + sinc(\frac{t}{T_s})T_s e^{j2\pi 21t}; \tag{1.7}$$

(refer Fig. 1.4), we get two sinusoidal components with frequencies 19 and 21 Hz. Figure 1.5 demonstrates the reconstruction of three different single tone signals using three different filters from the identical discrete sequence. But if we have two narrow Band pass filters that allow 11 Hz ($\frac{1}{2}sinc(\frac{t}{10T_s})e^{-j2\pi 11t} + sinc(\frac{t}{10T_s})e^{j2\pi 11t}$),

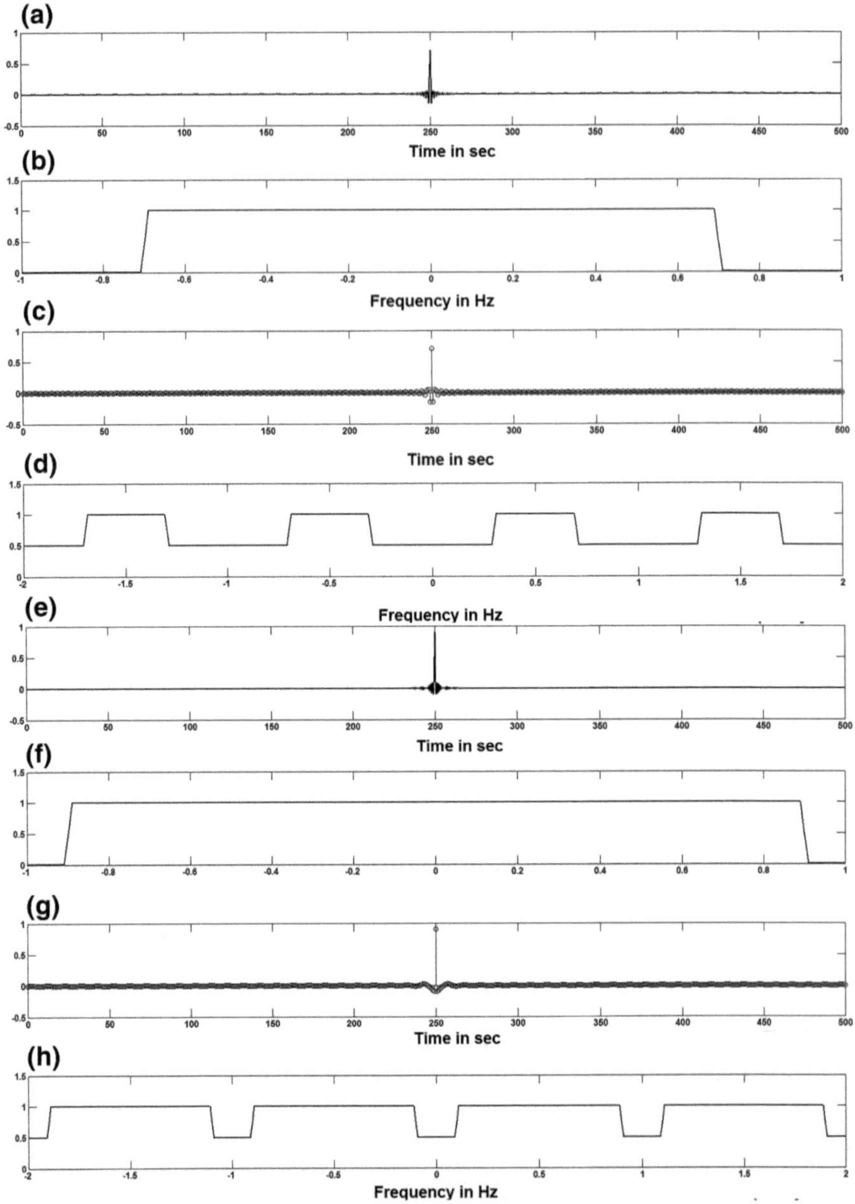

Fig. 1.11 Demonstrating Overlapping of spectrum in DTFT domain when $F_{max} > \frac{F_s}{2}$. **a** Low pass signal with the maximum frequency content of 0.7 Hz in time domain. **b** Corresponding spectrum. **c** Discrete version of the signal in (**a**) with $F_s = 1$ Hz. **d** Corresponding spectrum obtained using DTFT. **e** Low pass signal with the maximum frequency content of 0.9 Hz in time domain. **f** Corresponding spectrum. **g** Discrete version of the signal in (**e**) with $F_s = 1$ Hz. **h** Corresponding spectrum using DTFT

21 Hz ($\frac{1}{2}sinc(\frac{t}{10T_s})e^{-j2\pi 11t} + sinc(\frac{t}{10T_s})e^{j2\pi 11t}$) frequency, we get the corresponding sinusoidal components as the output of the filters (refer Figs. 1.6, 1.7 and 1.8). Ideally we would like to use the analog filter to reconstruct the samples back to the continuous domain. But to demonstrate this, we simulate reconstruction in discrete domain. The resolution to represent the impulse response of the analog filter is 1/1000 s. It is noted that it is not the actual sampling frequency used to sample the signal to get the discrete sequence (as described above). The spectrum of the low pass, bandpass filters used to demonstrate the reconstruction involves the periodic spectrum of 1000 Hz. This is due to discrete filter used to reconstruct the signal back. We reconstruct the signal back with the resolution 1/1000 s. In reality, the reconstruction is done using the analog filter to obtain the continuous signal.

```
function [res]=demonstratereconstruction(option)
%option=1: Using the  band pass filters constructed
%using the low pass reconstruction filter
%opition=2: Using 10 times narrow band pass filters
figure
subplot(2,1,1)
t=-2:1/1000:2;
x1=sin(2*pi*1*t);
plot(t,x1,'r')
hold on
x2=sin(2*pi*11*t);
plot(t,x2,'b')
hold on
x3=sin(2*pi*21*t);
plot(t,x3,'k')
Ts=1/10;
n=-20:1:20;
hold on
xn1=sin(2*pi*1*n*Ts);
plot(n*Ts,sin(2*pi*1*n*Ts),'m*')
subplot(2,1,2)
stem(n*Ts,sin(2*pi*1*n*Ts),'m*')


switch option
case 1
%Reconstruction using low pass filter
%Reconstruction using band pass filter 1
%Reconstruction using band pass filter 2
t1=-2:1/1000:2;
h1=sinc(t1/Ts);
w=window(@hamming,length(h1));
h1=h1.*w';
h2=(sinc(t1/Ts).*exp(-j*2*pi*11*t1)+sinc(t1/Ts).* ...
exp(j*2*pi*11*t1))/2;
h2=h2.*w';
h3=(sinc(t1/Ts).*exp(-j*2*pi*21*t1)+sinc(t1/Ts).*...
exp(j*2*pi*21*t1))/2;
h3=h3.*w';
figure
subplot(2,1,1)
plot(-2:1/1000:2,h1)
subplot(2,1,2)
```

```
plot(linspace(0,1000,length(h1)),abs(fft(h1)))
%%%%%%%%%%%%%%%%%%%%%%%%%%%%%%%%%%%%%%%%%%%%%%%%%%%%%%%%%%%%%%%
%OR%
case 2
%%%%%%%%%%%%%%%%%%%%%%%%%%%%%%%%%%%%%%%%%%%%%%%%%%%%%%%%%%%%%%%
%Reconstruction using low pass filter
%Reconstruction using 10 times narrow band pass filter 1
%Reconstruction using 10 times narrow band pass filter 2
t1=-2:1/1000:2;
h1=sinc(t1/Ts);
w=window(@hamming,length(h1));
h1=h1.*w';
h2=(sinc(t1/10*Ts).*exp(-j*2*pi*11*t1)+sinc(t1/10*Ts).* ...
exp(j*2*pi*11*t1))/2;
h2=h2.*w';
h3=(sinc(t1/10*Ts).*exp(-j*2*pi*21*t1)+sinc(t1/10*Ts).* ...
exp(j*2*pi*21*t1))/2;
h3=h3.*w';
figure
subplot(2,1,1)
plot(-2:1/1000:2,h1)
subplot(2,1,2)
plot(linspace(0,1000,length(h1)),abs(fft(h1)))
%%%%%%%%%%%%%%%%%%%%%%%%%%%%%%%%%%%%%%%%%%%%%%%%%%%%%%%%%%%%%%%
end

figure
subplot(2,1,1)
plot(-2:1/1000:2,h2)
subplot(2,1,2)
plot(linspace(0,1000,length(h2)),abs(fft(h2)))
figure
subplot(2,1,1)
plot(-2:1/1000:2,h3)
subplot(2,1,2)
plot(linspace(0,1000,length(h3)),abs(fft(h3)))

%Inserting zeros
temp=[];
for i=1:1:41
temp=[temp xn1(i) zeros(1,99)];
end
temp=temp(1:1:4001);
temp=temp.*window(@hamming,length(temp))';

y1=conv(temp,h1);
y2=conv(temp,h2);
y3=conv(temp,h3);
figure
subplot(2,3,1)
plot([-4:1/1000:4],y1,'b')
subplot(2,3,2)
plot([-4:1/1000:4],y2,'b')
subplot(2,3,3)
plot([-4:1/1000:4],y3,'b')
temp1=[];
temp2=[];
temp3=[];
t=-4+(1/1000):(1/1000):4-(1/1000);
```

```
%Trapezoidal rule
temp1=[];
temp2=[];
for f=0:1/231:50
temp1= [temp1 abs(y1(1)*(exp(-j*2*pi*f*(-4))+ ...
sum(y1(2:1:length(y1)-1).*exp(-j*2*pi*f*t))+ ...
y1(length(y1))*exp(-j*2*pi*f*(4))))];
temp2=[temp2 abs(y2(1)*(exp(-j*2*pi*f*(-4))+ ...
sum(y2(2:1:length(y2)-1).*exp(-j*2*pi*f*t))+ ...
y2(length(y2))*exp(-j*2*pi*f*(4))))];
temp3=[temp3 abs(y3(1)*(exp(-j*2*pi*f*(-4))+ ...
sum(y3(2:1:length(y3)-1).*exp(-j*2*pi*f*t))+ ...
y3(length(y3))*exp(-j*2*pi*f*(4))))];
end
f=0:1/231:50;
subplot(2,3,4)
plot(f, temp1)
subplot(2,3,5)
plot(f, temp2)
subplot(2,3,6)
plot(f,temp3)
```

1.2 Fourier Series (FS)

Any periodic signal (satisfying Dirichlet conditions) $x(t)$ with time period $T_0 = \frac{1}{f_0}$ (f_0 is the fundamental frequency) can be represented as the summation of sinusoid, cosines, and constant amplitude signal as follows.

$$x(t) = a_0 + \sum_{k=1}^{k=\infty} a_k sin(2\pi kf_0t) + \sum_{k=1}^{k=\infty} b_k cos(2\pi kf_0t) \tag{1.8}$$

As the signal is periodic, we consider the signal for one time period T_0 (represented as $x_T(t)$) as the vector. Also, we represent the set of sine and cosine signals in (1.8) for one period as follows.

$$S_{T,k}(t) = sin(2\pi kf_0t),\; C_{T,k}(t) = cos(2\pi kf_0t) \tag{1.9}$$

The vector $x_T(t)$ is viewed as the vector in the space spanned by the orthonormal basis vectors $\frac{2}{T_0}S_{T,k}(t)$, $\frac{2}{T_0}C_{T,k}(t)$ and $\frac{1}{T_0}$ (constant over the period 0 to T_0) and we obtain the co-efficients a_0, a_k and b_k ($k = 1, 2 \cdots$) as follows.

$$a_0 = \frac{1}{T_0}\int_0^{T_0} x_T(t)dt = \frac{1}{T_0}\int_0^{T_0} x(t)dt \tag{1.10}$$

$$a_k = \frac{2}{T_0}\int_0^{T_0} x_T(t)S_{T,k}(t)dt = \frac{2}{T_0}\int_0^{T_0} x(t)sin(2\pi kf_0t)dt \tag{1.11}$$

$$b_k = \frac{2}{T_0}\int_0^{T_0} x_T(t)C_{T,k}(t)dt = \frac{2}{T_0}\int_0^{T_0} x(t)cos(2\pi kf_0t)dt \quad (1.12)$$

1.2.1 Dirichlet Conditions

- $x(t)$ is absolutely integrable.
- $x(t)$ should have finite number of extrema in any given bounded interval.
- $x(t)$ must have finite number of discontinuities in any given bounded interval.

In this case, the signal in time domain is continuous and periodic and the frequency domain is discrete.

1.3 Fourier Transformation (FT)

Consider the aperiodic signal $x(t)$. It is assumed that the signal $x_{T_0}(t)$ is periodic with period $T_0 = \frac{1}{f_0}$ and $\lim_{T_0\to\infty} x_{T_0}(t) = x(t)$. Rewriting the Fourier series to represent the signal $x_{T_0}(t)$ as follows.

$$x_{T_0}(t) = \sum_{k=-\infty}^{k=\infty} c_k e^{j2\pi kf_0t} \quad (1.13)$$

The co-efficient c_k is obtained as follows.

$$c_k = \frac{1}{T_0}\int_{\frac{-T_0}{2}}^{\frac{T_0}{2}} x_{T_0}(t)e^{-j2\pi kf_0t}dt \quad (1.14)$$

Let $c_kT_0 = X(kf_0)$
Rewriting (1.13) using $X(kf_0)$, we get the following.

$$x_{T_0}(t) = \frac{1}{T_0}\sum_{k=-\infty}^{k=\infty} X(kf_0)e^{j2\pi kf_0t} \quad (1.15)$$

By applying the limit $\lim_{T_0\to\infty}$ on both sides of (1.15), we get the following.

$$x(t) = \int_{-\infty}^{\infty} X(f)e^{j2\pi ft}df \quad (1.16)$$

(1.16) implies,

$$X(f) = \int_{-\infty}^{\infty} x(t)e^{-j2\pi ft}dt \tag{1.17}$$

In this case, time domain is non-periodic and continuous.

1.4 Discrete Time Fourier Transformation (DTFT) and Sampling Theorem

The continuous time domain signal is sampled to obtain the discrete sequence. Based on the discussion in the Sect. (1.1), we understand that the corresponding frequency domain is periodic with the period of sampling frequency F_s. We also see that the amplitude of the frequency domain obtained with sampling is scaled by the factor F_s. Thus in the case of DTFT, time domain is discrete and the frequency domain is continuous and periodic. The DTFT pair is as given below.

$$X(f) = \sum_{n=-\infty}^{n=\infty} x(nT_s)e^{-j2\pi fnT_s} \tag{1.18}$$

$$x(nT_s) = x(n) = \frac{1}{F_s}\int_{-\frac{F_s}{2}}^{\frac{F_s}{2}} X(f)e^{j2\pi fnT_s}df \tag{1.19}$$

The spectrum (DTFT) of the discrete sequences sampled with the sampling frequency of 1 Hz from the continous signal whose maximum frequency is gradually increased from 0.1 Hz to 0.9 Hz is illustrated in the Figs. 1.9, 1.10, 1.11. It is seen in Fig. 1.9, there is no overlapping of spectrum occur in the case when $F_{max} < \frac{F_s}{2}$. Figure 1.10 illustrates the spectrum when $F_{max} = \frac{F_s}{2}$ and in Fig. 1.11 illustrates the overlapping of one period of the spectrum with neighbor occur that leads to the aliased component. This indicate that when the sampling frequency is chosen greater than or equal to twice the maximum frequency content of the signal, overlapping of one period of the spectrum does not occur (Aliased component will not get introduced) (Figs. 1.9, 1.10, 1.11).

Let the Fourier transformation of the signal $x(t)$ be represented as $X(f)$ and is zero if $|f| > \frac{W}{2}$. If the signal $x(t)$ is sampled with the sampling frequency F_s, we get the spectrum that is periodic with F_s. From the Fig. 1.11, it is observed that if

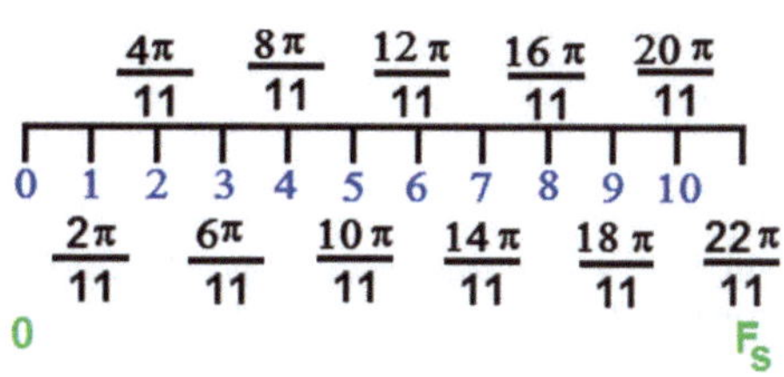

Fig. 1.12 Illustration of sampling points between 0 to 2π in DTFT domain to obtain DFT

$F_s < 2\frac{W}{2}$, overlapping of one period of the spectrum with the neighbor period occur. This is circumvented by choosing $F_s \geq 2\frac{W}{2}$. This is known as sampling theorem. It is possible to reconstruct the signal $x(t)$ from the discrete samples $x(nT_s)$ by filtering

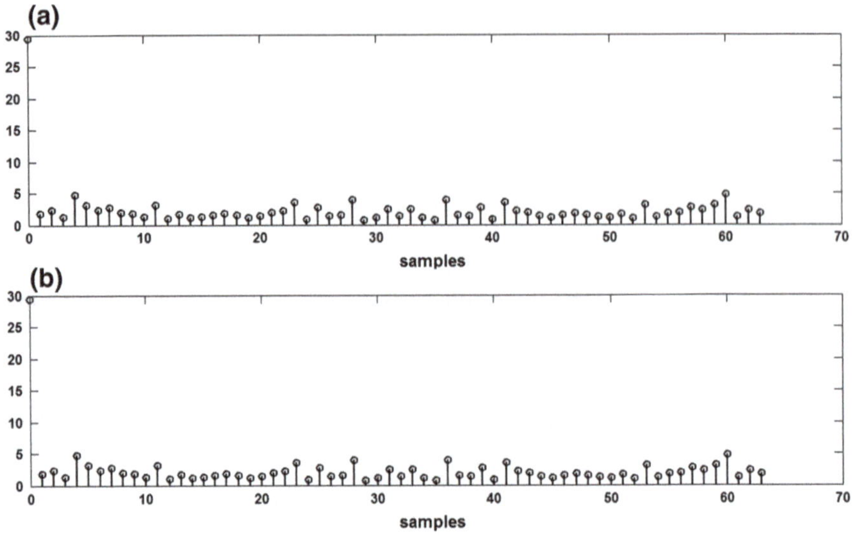

Fig. 1.13 **a** DFT computed using DIT. **b** DFT computed using the direct method (It is seen that both are identical)

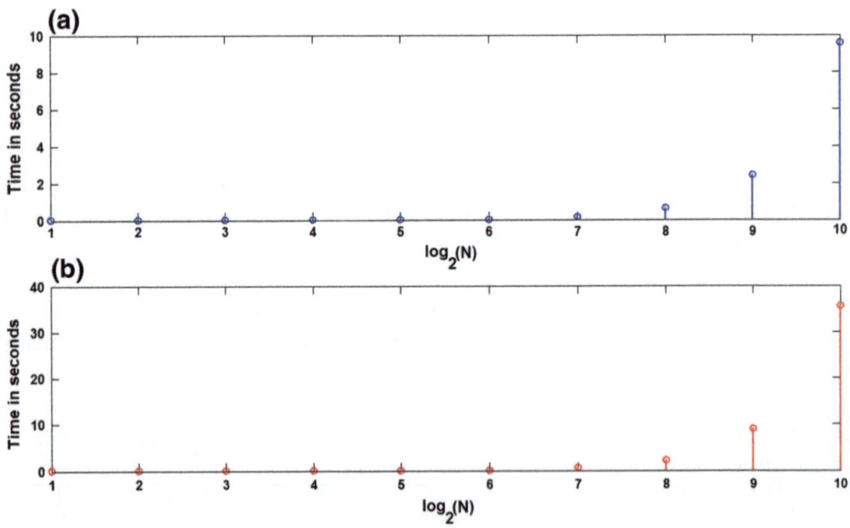

Fig. 1.14 **a** Time required to compute DFT using DIT method. **b** Time required to compute DFT using the direct method as the function of N

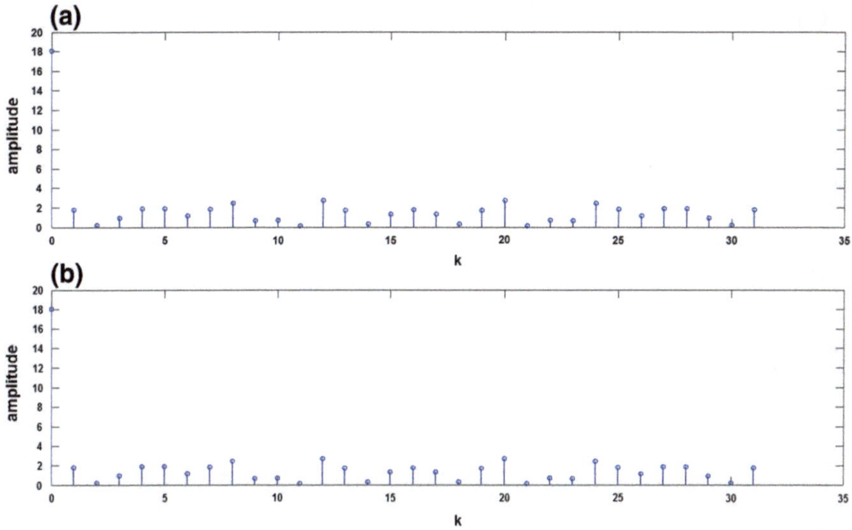

Fig. 1.15 **a** DFT computed using DIF. **b** DFT computed using the direct method (It is seen that both are identical)

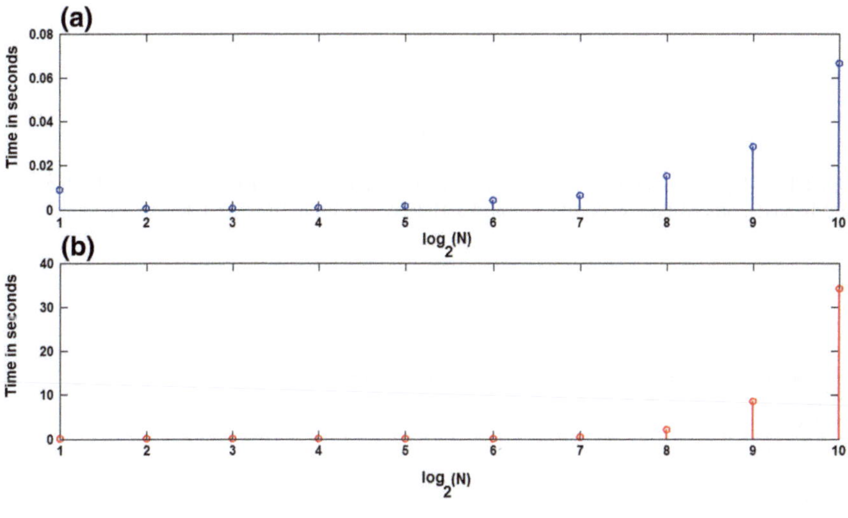

Fig. 1.16 **a** Time required to compute DFT using DIF method. **b** Time required to compute DFT using the direct method as the function of N

the discrete samples through the ideal low pass filer with cutoff frequency $= \frac{F_s}{2}$. This is equivalently represented in time domain (using sinc function based interpolation) as follows.

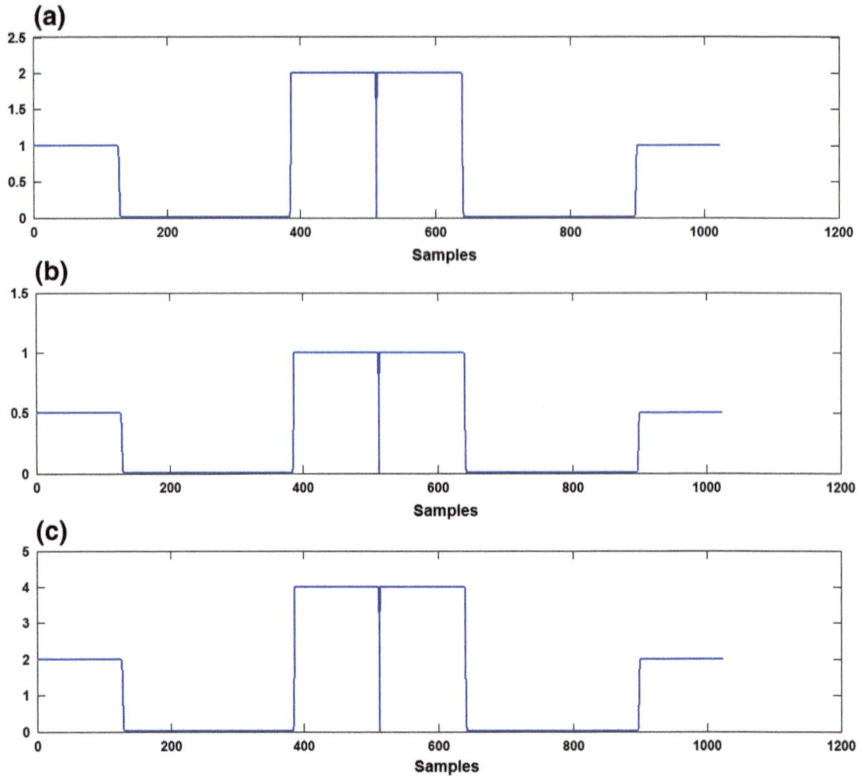

Fig. 1.17 Magnitude response computed using 1024-DFT using. **a** Direct method. **b** One level Sub-band decomposition. **c** Two-level Sub-band decomposition. (It is seen that all are identical)

```
%samplingtheorem.m
Demonstration of non-overlapping of spectrum in DTFT
m=1;
for cutoff=0.1:0.2:0.9
figure(m)
h=fir1(1000,cutoff);
Fs=2;
Ts=1/Fs;
%Assuming fmax=1 Hz
%(satisfying the sampling theorem)
n=0:1:length(h)-1;
subplot(4,1,1)
plot(n*Ts,h)
subplot(4,1,2)
m1=[];
for f=-6*Fs:0.01:6*Fs
    m1=[m1 abs(sum(h.*exp(-j*2*pi*f*n*Ts)))];
end
f=-6*Fs:0.01:6*Fs;
```

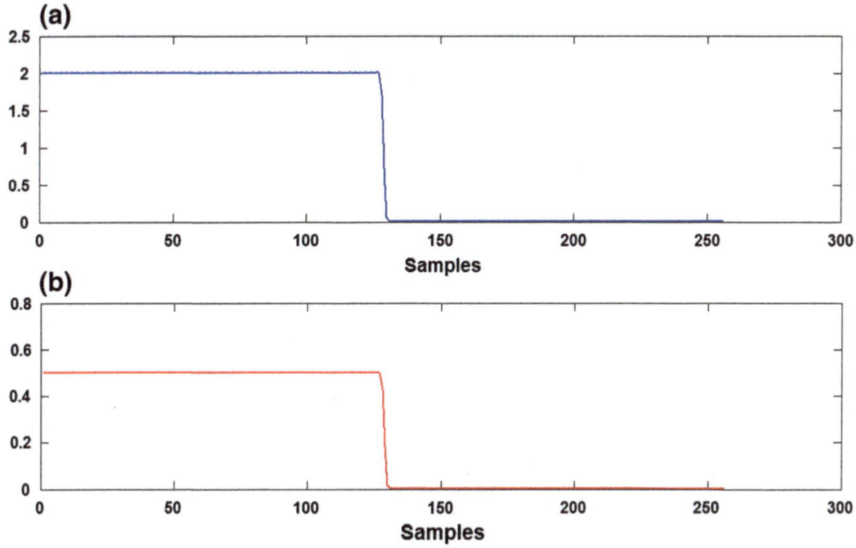

Fig. 1.18 Magnitude response: **a** 256 points of 1024-point DFT computed (0.2514 s) using the direct method. **b** 256-points of 1024-point DFT computed (0.0057 s) using Two-level sub-band decomposition

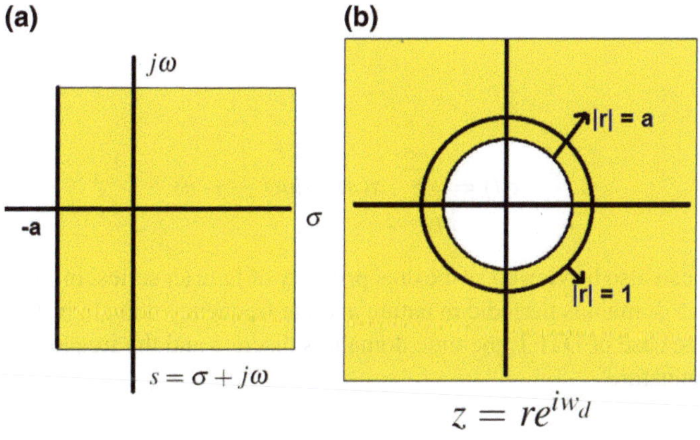

Fig. 1.19 **a** s-Plane with the shaded ROC. **b** z-Plane with the shaded ROC

```
plot(f,m1)
Ts1=2*Ts;
Fs1=Fs/2;
h1=h(1:2:1001)
n=0:1:length(h1)-1;
subplot(4,1,3)
stem(n*Ts1,h1)
m2=[];
for f=-2*6*Fs1:0.01:2*6*Fs1
    m2=[m2 abs(sum(h1.*exp(-j*2*pi*f*n*Ts1)))];
```

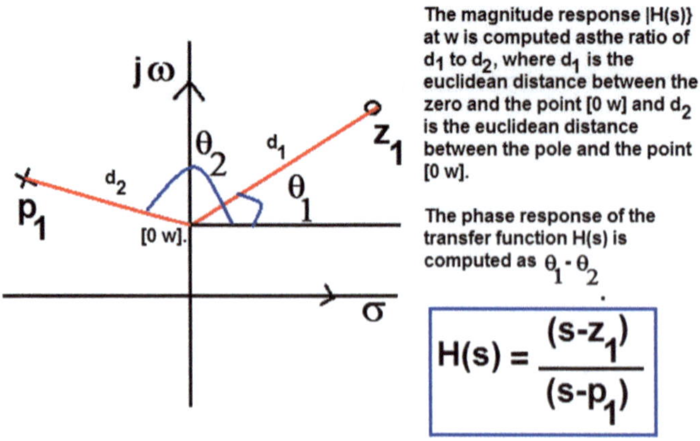

Fig. 1.20 Illustration of the Geometrical method of computing magnitude and phase response of the continous system transfer function $H(s)$

```
end
Fs1=1/Ts1;
f=-2*6*Fs1:0.01:2*6*Fs1;
%(Not satisfying the sampling theorem)
subplot(4,1,4)
plot(f,m2)
m=m+1;
end
```

$$x(t) = \sum_{n=-\infty}^{n=\infty} x(nT_s) sinc(\frac{t}{T_s} - n) \quad (1.20)$$

DTFT can also be viewed as the dual property of Fourier series. In case of Fourier series, time domain is periodic in nature and the frequency domain is discrete. Similarly in the case of DTFT, the time domain is discrete and the frequency domain is periodic in nature.

1.5 Discrete Fourier Transformation (DFT) (Sampling the Frequency Domain)

The spectrum of the discrete samples (DTFT) is periodic with sampling frequency F_s. DFT is obtained by sampling the DTFT for one period. Let us assume $N = 11$ number of samples are obtained from one period of the frequency domain DTFT. DTFT is represented as shown below.

$$X(e^{jw}) = \sum_{k=-\infty}^{k=\infty} x(n)e^{-jwn} \tag{1.21}$$

DFT sequence is obtained by sampling $X(e^{jw})$ for the duration $0 \leq w \leq 2\pi$ once in $2\frac{\pi}{N}$ and is represented as follows.

$$X(k) = \sum_{n=0}^{n=N-1} x(n)e^{-j\frac{2\pi}{N}nk} \tag{1.22}$$

The sequence $x(n)$ is obtained from the sequence $X(k)$ as $x(n) = \frac{1}{N}\sum_{k=0}^{k=N-1} X(k)e^{j\frac{2\pi}{N}nk}$ and the proof is as shown below. Substitute (1.22) in $x(n) = \frac{1}{N}\sum_{k=0}^{k=N-1} X(k)e^{j\frac{2\pi}{N}nk}$, we get the following.

$$\begin{aligned}
&\frac{1}{N}\sum_{k=0}^{k=N-1}\sum_{l=0}^{l=N-1} x(l)e^{-j\frac{2\pi}{N}lk}e^{j\frac{2\pi}{N}nk} \\
&= \frac{1}{N}\sum_{l=0}^{l=N-1} x(l)\sum_{k=0}^{k=N-1} e^{-j\frac{2\pi}{N}lk}e^{j\frac{2\pi}{N}nk} \\
&= \frac{1}{N}\sum_{l=0}^{l=N-1} x(l)\sum_{k=0}^{k=N-1} e^{-j\frac{2\pi}{N}(l-n)k} \\
&= \frac{1}{N}x(0)\sum_{k=0}^{k=N-1} e^{-j\frac{2\pi}{N}(0-n)k} + x(1)\sum_{k=0}^{k=N-1} e^{-j\frac{2\pi}{N}(1-n)k} + \cdots + \\
&\quad x(n)\sum_{k=0}^{k=N-1} e^{-j\frac{2\pi}{N}(n-n)k} + \cdots \\
&\quad + x(N-1)\sum_{k=0}^{k=N-1} e^{-j\frac{2\pi}{N}(N-1-n)k} \\
&\Rightarrow x(n) = 0 + 0 + \cdots + Nx(n) + 0 + \cdots + 0 \\
&\quad = x(n)
\end{aligned}$$

Hence proved. In this case, both time domain and frequency domain are discrete.

1.5.1 Vector Space Interpretation of DFT

The typical DFT matrix D_N for arbitrary N is as shown below.

$$\begin{pmatrix} 1 & 1 & 1 & \cdots & 1 \\ 1 & W_N^{1*1} & W_N^{1*2} & \cdots & W_N^{1*(N-1)} \\ 1 & W_N^{2*1} & W_N^{2*2} & \cdots & W_N^{2*(N-1)} \\ 1 & W_N^{3*1} & W_N^{3*2} & \cdots & W_N^{3*(N-1)} \\ \cdots & \cdots & \cdots & \cdots & \cdots \\ 1 & W_N^{(N-1)*1} & W_N^{(N-1)*2} & \cdots & W_N^{(N-1)*(N-1)} \end{pmatrix}$$

where, $W_N = e^{\frac{j2\pi}{N}}$ is the twiddle factor. The columns of the DFT matrix D_N is orthogonal to each other, i.e., $D_N^H D_N = NI_N$, where I_N is the identity matrix of size $N \times N$. Also DFT matrix is the unitary matrix, i.e., $D_N^H D_N = D_N D_N^H = NI_N$. The arbitrary vector $v \in C^n$ is represented as the linear combinations of the columns of the DFT matrix, i.e., $v = \frac{1}{N} D_N u$. The corresponding co-efficients are obtained as $u = D_N^H v$. The vector u is known as DFT coefficients and are obtained using DFT. The method of obtaining v is known as IDFT. Properties of the DFT matrix is summarized below. The frequency in radians and in Hz corresponding to the sample number in DFT domain is illustrated in Fig. 1.12.

1. The columns of the scaled DFT matrix $\frac{1}{N}$ is orthogonal to each other. To make it as the orthonormal vectors, they are scaled with $\frac{1}{\sqrt{(N)}}$. Let it be $SD_N = \frac{1}{N}$
2. The eigenvectors of the DFT matrix takes the value i, $-i$ or 1
3. Circulant matrix (A (say)) is diagonalized using DFT matrix as $SD_N A SD_N^H = \frac{1}{N} D_N A D_N^H = D$, D is the diagonal matrix.
   This implies $A = \frac{1}{N} D_N^H D D_N$. The vector $y = Ax$ (A as the circulant matrix) is the circular convolution of the vector x and the first column of the matrix A and is computed as $\frac{1}{N} D_N^H D D_N x$. In this $D_N x$ is viewed as computing the DFT OF the vector x (say z_1). $DD_N x$ is viewed as the scaled version of the vector z_1, with the diagonal elements of the matrix D (say z_2). Further $\frac{1}{N} D_N^H D D_N x$ is viewed as the IDFT of the vector z_2. It is also observed that the diagonal elements of the matrix D is the DFT of the first column of the matrix A.
4. Demonstration with $A = \begin{pmatrix} 1 & 4 & 3 & 2 \\ 2 & 1 & 4 & 3 \\ 3 & 2 & 1 & 4 \\ 4 & 3 & 2 & 1 \end{pmatrix}$.
   $x = \begin{pmatrix} 5 & 6 & 7 & 8 \end{pmatrix}$ we get the following.

$$D_N = \begin{pmatrix} 1.0000 & 1.0000 & 1.0000 & 1.0000 \\ 1.0000 & 0.0000 + 1.0000i & -1.0000 + 0.0000i & -0.0000 - 1.0000i \\ 1.0000 & -1.0000 + 0.0000i & 1.0000 - 0.0000i & -1.0000 + 0.0000i \\ 1.0000 & -0.0000 - 1.0000i & -1.0000 + 0.0000i & 0.0000 + 1.0000i \end{pmatrix}$$

$$z1 = \begin{pmatrix} 1 & 4 & 3 & 2 \\ 2 & 1 & 4 & 3 \\ 3 & 2 & 1 & 4 \\ 4 & 3 & 2 & 1 \end{pmatrix}$$

$$D = \begin{pmatrix} 26.0000 & 0 & 0 & 0 \\ 0 & -2.0000 + 2.0000i & 0 & 0 \\ 0 & 0 & -2.0000 - 0.0000i & 0 \\ 0 & 0 & 0 & -2.0000 - 2.0000i \end{pmatrix}$$

$$z2 = \begin{pmatrix} 10.0000 \\ -2.0000 + 2.0000i \\ -2.0000 - 0.0000i \\ -2.0000 - 2.0000i \end{pmatrix}$$

$$z3 = \begin{pmatrix} 66.0000 + 0.0000i \\ 68.0000 + 0.0000i \\ 66.0000 + 0.0000i \\ 60.0000 - 0.0000i \end{pmatrix}$$

The $z3$ thus obtained is the circular convolution of the vector x and the first column of the matrix A, which is obtained as Ax.

```
%circconvusingDFT.m
%Circular convolution of a with b using DFT-IDFT
a=[1 2 3 4]';
b=[5 6 7 8]';
%Constructing DFT matrix
twf=exp(j*2*pi/4);
n=0:1:3;
k=[0:1:3]';
DN=[twf.^n; twf.^n; twf.^n; twf.^n];
DN=[DN(:,1).^k DN(:,2).^k DN(:,3).^k DN(:,4).^k]
%DFT of a
z1=DN'*a;
temp=DN'*b;
D=diag(temp);
z2=D*z1;
z3=(1/4)*DN*z2;
%Constructing circulant matrix
C=[1 2 3 4;4 1 2 3;3 4 1 2;2 3 4 1]';
%Circular convolution of two sequence
z3_direct=C*b;
```

1.5.2 *Fast Fourier Transformation*

The number of computations to compute DFT is reduced by the algorithm namely Fast Fourier Transformation as described below.

1.5.2.1 Decimation in Time for N-Point DFT

The DFT of the sequence $x(n)$ is decomposed as the summation of two terms as described below.

$$X(k) = \sum_{n=0}^{n=N-1} x(n)e^{\frac{-j2\pi nk}{N}} \tag{1.23}$$

$$= \sum_{n=0}^{n=\frac{N}{2}-1} x(2n)e^{\frac{-j\pi 2nk}{\frac{N}{2}}} + e^{\frac{-j2\pi k}{N}} \sum_{n=0}^{n=\frac{N}{2}-1} x(2n+1)e^{\frac{-j2\pi nk}{\frac{N}{2}}} \tag{1.24}$$

This is realized using recursive technique. Refer Figs. 1.13 and 1.14 for illustration.

```
%dit.m
function [res]=dit(x,N)
%Decimation in time
k=1;
if(N~=2)
part1=x(1:2:length(x));
part2=x(2:2:length(x));
N=N/2;
k=0:1:N-1;
res1=dit(part1,N)+dit(part2,N).*exp(-j*2*pi*k/(2*N));
res2=dit(part1,N)-dit(part2,N).*exp(-j*2*pi*k/(2*N));
res=[res1 res2];
else
res1=x(1)+x(2);
res2=x(1)-x(2);
res=[res1 res2];
end

%executedit
OPTIONN=[2 4 8 16 32 64 128 256 512 1024];
time1=[];
time2=[];
for i=1:length(OPTIONN)
N=OPTIONN(i);
x=rand(1,N);
tic
res=dif(x,N);
n=0:1:N-1;
c=dec2bin(n);
p=bin2dec([c(:,size(c,2):-1:1)]);
DIFR=res(p+1);
time1=[time1 toc];

tic
for k=0:1:length(x)-1
    sum=0;
    for n=0:1:length(x)-1
        sum=sum+x(n+1)*exp(-j*2*pi*n*k/N);
    end
RES(k+1)=sum ;
end
time2=[time2 toc];
end

figure
subplot(2,1,1)
stem(0:1:length(x)-1,abs(DIFR))
subplot(2,1,2)
```

```
stem(0:1:length(x)-1,abs(RES))

figure
stem(time1)
hold on
stem(time2,'r')


function [res]=dif(x,N)
Decimation in frequency
part1=x(1:1:(N/2));
part2=x((N/2)+1:1:N);
if(N~=2)
    n=0:1:(N/2)-1;
    temp1=part1+part2;
    temp2=(part1-part2).*exp(-j*2*pi*n/(N))
    N=N/2;
    part1=dif(temp1,N);
    part2=dif(temp2,N);
    res=[part1 part2]
else
    res(1)=part1+part2;
    res(2)=part1-part2;
end
```

1.5.2.2 Decimation in Frequency for N-Point DFT

The DFT of the sequence $x(n)$ is decomposed as the summation of two terms as described below.

$$X(k) = \sum_{n=0}^{n=(N-1)} x(n)e^{\frac{-j2\pi nk}{N}} \tag{1.25}$$

$$= \sum_{n=0}^{n=\frac{N}{2}-1} x(n)e^{\frac{-j2\pi nk}{N}} + \sum_{n=\frac{N}{2}}^{n=(N-1)} x(n)e^{\frac{-j2\pi nk}{N}} \tag{1.26}$$

Consider the second term $\sum_{n=\frac{N}{2}}^{n=N-1} x(n)e^{\frac{-j2\pi nk}{N}}$.

Let $m = n - \frac{N}{2}$, we get the following $\sum_{m=0}^{m=\frac{N}{2}-1} x(m+\frac{N}{2})e^{\frac{-j2\pi(m+\frac{N}{2})k}{N}}$. Thus $X(k)$ is computed as the following.

$$X(k) = \sum_{n=0}^{n=\frac{N}{2}-1} x(n)e^{\frac{-j2\pi nk}{N}} + e^{\frac{-j2\pi(N/2)k}{N}} \sum_{m=0}^{m=\frac{N}{2}-1} x(m+\frac{N}{2})e^{\frac{-j2\pi mk}{N}} \tag{1.27}$$

Replacing $k = 2k'$, we get the following.

$$X(2k') = \sum_{n=0}^{n=\frac{N}{2}-1} (x(n) + x(n + \frac{N}{2}))e^{\frac{-j2\pi nk'}{N/2}} \tag{1.28}$$

Similarly $k = 2k' + 1$, we get the following.

$$X(2k'+1) = \sum_{n=0}^{n=\frac{N}{2}-1} (x(n) - x(n + \frac{N}{2}))e^{\frac{-j2\pi nk'}{N/2}} e^{\frac{-j2\pi n}{N}} \tag{1.29}$$

This is realized using the recursive method (refer m-file). The order in which the DFT is obtained is in the bit reversal order, i.e., decimated in the frequency domain and hence this method is known as Decimation in frequency. Refer Figs. 1.15 and 1.16 for illustration.

```
%dif.m
Decimation in frequency
function [res]=dif(x,N)
part1=x(1:1:(N/2));
part2=x((N/2)+1:1:N);
if(N~=2)
    n=0:1:(N/2)-1;
    temp1=part1+part2;
    temp2=(part1-part2).*exp(-j*2*pi*n/(N))
    N=N/2;
    part1=dif(temp1,N);
    part2=dif(temp2,N);
    res=[part1 part2]
else
    res(1)=part1+part2;
    res(2)=part1-part2;
end
```

```
%executedif.m
OPTIONN=[2 4 8 16 32 64 128 256 512 1024];
time1=[];
time2=[];
for i=1:length(OPTIONN)
N=OPTIONN(i);
x=rand(1,N);
tic
res=dif(x,N);
n=0:1:N-1;
c=dec2bin(n);
p=bin2dec([c(:,size(c,2):-1:1)]);
DIFR=res(p+1);
time1=[time1 toc];
tic
for k=0:1:length(x)-1
    sum=0;
    for n=0:1:length(x)-1
        sum=sum+x(n+1)*exp(-j*2*pi*n*k/N);
    end
```

```
RES(k+1)=sum ;
end
time2=[time2 toc];
end

figure
subplot(2,1,1)
stem(0:1:length(x)-1,abs(DIFR))
subplot(2,1,2)
stem(0:1:length(x)-1,abs(RES))

figure
subplot(2,1,1)
stem(log2(OPTIONN),time1)
subplot(2,1,2)
stem(log2(OPTIONN),time2,'r')
```

1.5.3 Sub-band Discrete Fourier Transformation

Suppose the spectrum of the given sequence occupies only the particular band between 0 to 2π radians (corresponding to F_{max} Hz). It is not efficient to compute the DFT for the complete range between 0 to 2π. Sub-band DFT paves the way to compute the DFT in the intended region of the complete spectrum. This is illustrated with an example as follows.

- Let the given sequence be represented as $[X_0\ X_1\ X_2 \cdots X_{1023}]$ that occupies the spectrum ranging from 0 to $\frac{\pi}{2}$. $1024 - point$ DFT occupies the complete range of frequency from 0 to 2π. Sub-band DFT to compute DFT points related to the range 0 to $\frac{\pi}{4}$ is computed using two-level decomposition as follows.
- $g_1 = [X_0\ X_2 \cdots X_{1022}] + [X_1\ X_3 \cdots X_{1023}]$
- $g_2 = [X_0\ X_2 \cdots X_{1022}] - [X_1\ X_3 \cdots X_{1023}]$
- $g_{11} = [g_0\ g_2 \cdots g_{510}] + [g_1\ g_3 \cdots g_{510}]$
- $g_{12} = [g_0\ g_2 \cdots g_{510}] - [g_1\ g_3 \cdots g_{510}]$
- $g_{21} = [g_1\ g_3 \cdots g_{511}] + [g_1\ g_3 \cdots g_{511}]$
- $g_{22} = [g_1\ g_3 \cdots g_{511}] - [g_1\ g_3 \cdots g_{511}]$
- Compute DFT of the sequence g_{11} to obtain G_{11} Similarly construct Gij, for $i = 1, 2$ and $j = 1, 2$.
- The actual DFT sequence of the given signal over the region 0 to $\frac{\pi}{2}$ (256-point FFT) is computed as the following. $X(k) = [1\ e^{\frac{-j2\pi k}{1024}}\ e^{\frac{-j2\pi 2*k}{1024}}\ e^{\frac{-j3\pi 2*k}{1024}}]\frac{1}{4}[1111]G_{11}^T(K)$. This is computed for $k = 0$ to $k = 255$ to obtain the 256-point DFT.
- DFT points from 256 to 512 are computed as the following. $X(k) = [1\ e^{\frac{-j2\pi k}{1024}}\ e^{\frac{-j2\pi 2*k}{1024}}\ e^{\frac{-j3\pi 2*k}{1024}}]\frac{1}{4}[1 - 1\ 1 - 1]G_{21}^T(K)$ for $k = 0$ to $k = 255$
- Similarly DFT points from 513 to 768 (256-points) are computed respectively as the following. $X(k) = [1\ e^{\frac{-j2\pi k}{1024}}\ e^{\frac{-j2\pi 2*k}{1024}}\ e^{\frac{-j3\pi 2*k}{1024}}]\frac{1}{4}[1\ 1\ -1\ -1]G_{12}^T(K)$ for $k = 0$ to $k = 255$ and

- Similarly DFT points from 769 to 1023 (256-points) are computed respectively as the following. $X(k) = [1\ e^{\frac{-j2\pi k}{1024}}\ e^{\frac{-j2\pi 2*k}{1024}}\ e^{\frac{-j3\pi 2*k}{1024}}]\frac{1}{4}[1 -1\ -11]G_{22}^{T}(K)$ for $k = 0$ to $k = 255$
- The vectors [1 1 1 1], [1 $-$ 1 1 $-$ 1], [1 1 $-$ 1 $-$ 1], [1 $-$ 1 $-$ 1 1] (used in the computation) form the rows of the 4×4 hadamard matrix.
- Figure 1.17 illustrates the magnitude response of the DFT computed using FFT, Sub-band first-level decomposition, Sub-band second-level decomposition. Also Fig. 1.18 demonstrates that the first 256 samples of the 1024-point DFT computed using 1024-point FFT and two-level sub-band DFT are identical. It is also observed that the time required to compute first 256 samples of the 1024-point DFT are 0.2514 and 0.0057 s when the direct method and the sub-band DFT methods are used respectively.
- Thus Sub-band DFT helps in obtaining the intended parts of the DFT sequence without calculating all the samples (like FFT)
- This can also be generalized to obtain the higher level sub-band decomposition.

```
%subband.m
%Subband Discrete Fourier Transform
X=fir2(1023,linspace(0,1,1024),[2*ones(1,256) zeros(1,512) 4*ones(1,255) 0]);
[H,W]=freqz(X);
figure
plot(W,abs(H));
%First level representation
temp=reshape(X,2,512);
g1=(1/2)*(temp(1,:)+temp(2,:));
G1=fft(g1);
g2=(1/2)*(temp(1,:)-temp(2,:));
G2=fft(g2);
i=0:1:1;
TF=exp(-j*2*pi*i/1024)';
%Construct matrix
M=[];
for k=0:1:1023
M=[M TF.^k];
end
M=M';
G=[G1' G2']';
G=repmat(G,1,2);
RES=M*inv(hadamard(2))*G;
figure(1)
subplot(2,1,1)
plot(abs(diag(RES)))
subplot(2,1,2)
plot(abs(fft(X)))

%second level representation
temp1=reshape(g1,2,256);
temp2=reshape(g2,2,256);
g11=(1/2)*(temp1(1,:)+temp1(2,:));
G11=fft(g11);
g12=(1/2)*(temp1(1,:)-temp1(2,:));
G12=fft(g12);
g21=(1/2)*(temp2(1,:)+temp2(2,:));
```

```
G21=fft(g21);
g22=(1/2)*(temp2(1,:)-temp2(2,:));
G22=fft(g22);
%Construction of DFT matrix
i=0:1:3;
TF=exp(-j*2*pi*i/1024)';
%Construct matrix
M=[];
for k=0:1:1023
M=[M TF.^k];
end
M=M';
G=[G11' G21' G12' G22']';
G=repmat(G,1,4);
RES=M*inv(hadamard(4))*G;
figure(2)
subplot(2,1,1)
plot(abs(diag(RES)))
subplot(2,1,2)
plot(abs(fft(X)))

tic
N=1024;
for k=0:1:255
s=0;
for n=0:1:1023
s=s+X(n+1)*exp(-j*2*pi*n*k/N);
end
res1(k+1)=s;
end
time1=toc
tic
RES=M(1:1:256,:)*inv(hadamard(4))*G(:,1:1:256);
res2=diag(RES);
time2=toc
figure
subplot(2,1,1)
plot(abs(res1))
subplot(2,1,2)
plot(abs(res2),'r')
```

1.6 Continous System

- Impulse response of the system: The response of the system to the impulse signal as the input is the impulse response $h(t)$ of the system. The impulse signal $\delta(t)$for the continuous case is as described below. $\int_{-\infty}^{\infty} \delta(t)dt = \int_{-\Delta}^{\Delta} dt = 1$ for every values of $\Delta > \varepsilon > 0$. The sifting property of the impulse is given as the following.

$$\int_{-\infty}^{\infty} f(t)\delta(t-t_0)dt = f(t_0)\int_{t_0^-}^{t_0^+} \delta(t-t_0)dt = f(t_0) \tag{1.30}$$

- Response to the eigensignal to the linear system: If the signal $e^{j2\pi f_0 t}$ is the input to the linear system with impulse response $h(t)$ (with transfer function $H(f)$, the output of the system is given as $e^{j2\pi f_0 t}H(f_0)$. The generalized input signal $x(t)$ is represented using Inverse Fourier Transformation as $\int_{-\infty}^{\infty} X(f)e^{j2\pi ft}df$ and hence the corresponding output signal $y(t)$ is given as $y(t) = \int_{-\infty}^{\infty} X(f)H(f)e^{j2\pi ft}df$. Hence $Y(f) = X(f)H(f)$. By taking IFT on both sides, we obtain, $y(t) = \int_{-\infty}^{\infty} h(\tau)x(t-\tau)d\tau$).This is defined as the convolution of $x(t)$ with $h(t)$
- Linearity: Let $x_i(t)$ be the input to the system and the corresponding output of the system is $y_i(t)$. The system is said to be linear if the output of the system is $\sum_{i=1}^{N} a_i y_i(t)$ corresponding to $\sum_{i=1}^{N} a_i x_i(t)$ as the input to the system.
- Causality: Two arbitrary inputs $x_1(t)$ and $x_2(t)$ are identical, i.e., $x_1(t) = x_2(t)$ for $t \leq t_0$, then the corresponding outputs satisfy $y_1(t) = y_2(t)$ for $t \leq t_0$ for all x_1, x_2 and t_0. The necessary and sufficient condition for system to be causal is $h(t) = 0$ for $t < 0$.
Proof:Consider the impulse response of the system be $h(t)$. Let $x_1(t)$ and $x_2(t)$ are identical for $t \leq t_0$. For $t \leq t_0$

$$y_1(t_0) = \int_{-\infty}^{t_0} h(\tau)x_1(t-\tau)d\tau$$
$$\Rightarrow y_1(t) = \int_{0}^{t_0} h(\tau)x_1(t-\tau)d\tau$$
$$y_2(t_0) = \int_{-\infty}^{t_0} h(\tau)x_2(t-\tau)d\tau$$
$$\Rightarrow y_2(t) = \int_{0}^{t_0} h(\tau)x_1(t-\tau)d\tau$$
$$\Rightarrow y_1(t_0) = y_2(t_0)$$

Thus if $h(t) = 0$ for $t \leq 0$, the system is the causal system. Consider the system with $h(-5) \neq 0$ and let $x_1(t) = 0\ \forall\ t$ and $x_2(t) = \delta(t-5)$. It is observed that the values of $x_1(t)$ and $x_2(t)$ are identical upto $t \leq 0$, the values for $y_1(0)$ and $y_2(0)$ are computed as follows.

$$y_1(0) = \int_{-\infty}^{\infty} x_1(-\tau)h(\tau)d(\tau) = 0$$
$$y_2(0) = \int_{-\infty}^{\infty} x_2(-\tau)h(\tau)d(\tau) = x_2(5)$$

$y_1(0) \neq y_2(0)$ and hence $h(t)$ is causal only if $h(t) = 0$ for $t \leq 0$. Hence proved.
- Stability: A signal is bounded if its magnitude is limited to a finite value at all times. If the bounded input gives bounded output (BIBO), it is the stable system. The system with impulse response $h(t)$ is BIBO stable if and only

$$\int_{-\infty}^{\infty} |h(\tau)|d\tau \leq \infty \tag{1.31}$$

Proof: Using Schwartz inequality,

$$y(t) = \int_{-\infty}^{\infty} h(\tau)x(t-\tau)d\tau$$
$$|y(t)| \leq \int_{-\infty}^{\infty} |h(\tau)|d\tau \int_{-\infty}^{\infty} x(t-\tau)|d\tau$$
$$\leq K \int_{-\infty}^{\infty} |h(\tau)|d\tau$$

Thus if $\int_{-\infty}^{\infty} |h(\tau)|d\tau \leq \infty$, $|y(t)| \leq \infty$, i.e., stable system.
Suppose that $h(t)$ is not satisfying $\int_{-\infty}^{\infty} |h(\tau)|d\tau \leq \infty$ and let for instance, there exists some finite t_0 such that $\int_{-\infty}^{t_0} |h(\tau)|d\tau = \infty$. Consider the case the input satisfy $x(t-\tau) = -1$ for $h(\tau) \geq 0$ and $x(t-\tau) = -1$ for $h(\tau) \leq 0$. This implies the following.

$$y(t_0) = \int_{-\infty}^{t_0} h(\tau)x(t-\tau)d\tau = \int_{-\infty}^{t_0} |h(\tau)|d\tau = \infty$$

Thus the system is stable only if $\int_{-\infty}^{\infty} |h(\tau)|d\tau \leq \infty$. Hence proved.

- Time in-variant system: Let $y(t)$ is the output corresponds to the signal $x(t)$, then the system is said to time in-variant if $y(t-t_0)$ is the output corresponds to the input $x(t-t_0)$
- Memory system: If the output of the system $y(t)$ depends on $x(t+t_0)$ and/or $y(t+t_0)$, with $t_0 = 0$, then the system is said to memory-less system.
- Linear phase of the system: Let the transfer function of the system is represented as $|H(f)|e^{j\phi(f)}$. The magnitude response is given as $|H(f)|$ and the phase response is given as $\phi(f)$, Let the output signal corresponding to the input signal $x(t) - a_1e^{j2\pi f_1 t|} + a_2e^{j2\pi f_2 t|}$ is obtained as $a_1e^{j2\pi f_1 t}|H(f_1)|e^{j\phi(f_1)} + a_2e^{j2\pi f_2 t}|H(f_2)|e^{j\phi(f_2)}$ respectively. Suppose the input signal is delayed by t_0, the corresponding output is given as the following.

$$a_1e^{j2\pi f_1(t-t_0)}|H(f_1)|e^{j\phi(f_1)} + a_2e^{j2\pi f_2(t-t_0)}|H(f_2)|e^{j\phi(f_2)}$$

which may be equal to $y(t-t_0)$. But suppose if we choose $\phi(f) = -2\pi\tau f$, where τ is constant, we get the output as the following.

$$a_1e^{j2\pi f_1(t-t_0)}|H(f_1)|e^{j\phi(f_1)} + a_2e^{j2\pi f_2(t-t_0)}|H(f_2)|e^{j\phi(f_2)}$$
$$a_1e^{j2\pi f_1(t-t_0)}|H(f_1)|e^{-j2\pi\tau f_1} + a_2e^{j2\pi f_2(t-t_0)}|H(f_2)|e^{j2\pi\tau f_2}$$
$$a_1e^{j2\pi f_1(t-t_0-\tau)}|H(f_1)| + a_2e^{j2\pi f_2(t-t_0-\tau)}|H(f_2)|$$
$$y(t-t_0)$$

Thus if the phase response is linear, we get undistorted delayed version of the output signal if the linear combinations of the delayed eigensignals are given as the input to the linear system.

1.7 Laplace Transformation (s-Transformation)

The Laplace transformation of the right-sided signal $x(t)$, i.e. $x(t) = 0$ for $t < 0$ is given as

$$X(s) = \int_{0-}^{\infty} x(t)e^{-st}dtX(s) = \int_{0-}^{\infty} x(t)e^{-\sigma t}e^{-j\omega t}dt \tag{1.32}$$

where $s = \sigma + j\omega$. The set of all values of s for which $X(s)$ (for the particular $x(t)$) is finite is the region of convergence (ROC). It helps in computing the inverse laplace transformation. For instance consider $x(t) = e^{-at}u(t)$, where $u(t)$ is the unit step function. we compute the Laplace transformation as shown below.

$$X(s) = \int_{0}^{\infty} e^{-at}e^{-\sigma t}e^{-j\omega t}dt = \int_{0}^{\infty} e^{-(a+\sigma+j\omega)t}dt = \frac{e^{-(a+\sigma+j\omega)\infty}}{-(a+\sigma+j\omega)} - \frac{e^{-(a+\sigma+j\omega)0}}{-(a+\sigma+j\omega)} \tag{1.33}$$

The value of the first term is zero if $a + \sigma > 0$ and ∞ if $a + \sigma < 0$ and not defined for $a + \sigma = 0$. Hence to make X(s) to exist, we assume $a + \sigma > 0$ and compute the Laplace transformation as the following.

$$X(s) = \frac{1}{s+a} \tag{1.34}$$

Thus the value of $X(s)$ is finite if $a + \sigma > 0$. This is the ROC. (refer Fig. 1.19)

The s-plane is the complex plane with each point represents the complex number represented in the rectangular form. It is observed that if $\sigma = 0$, Laplace transformation becomes Fourier transformation. Hence if $X(s)$ is finite for all values of ω with $\sigma = 0$, we conclude that Fourier transformation exists. In other words, Fourier transformation exists if the ROC includes the $j\omega$ axis. Thus Fourier tansformation can be viewed as the special case of Laplace transformation with $\sigma = 0$. The set of all values of s for which $X(s) = 0$ are the zeros of the laplace transformation $X(s)$. Similarly, the set of all values of s for which $X(s) = \infty$ are the poles of the $X(s)$.

1.7.1 Properties of the ROC of the Laplace Transformation for the Right Sided Sequence

1. ROC is of the infinite length vertical rectangular strip shape in the s-plane.

2. For the Right Sided Sequence (R.S.S), ROC is in the right side of the vertical strip passes through the poles of the $H(s)$.
3. Hence to include jw axis in the ROC, all the poles of the Laplace transformation should lie in the left side of the jw axis. In this context, the transfer function $H(s)$ is said to be stable if all the poles lie in the left of the jw axis
4. ROC does not contain poles of $H(s)$.
5. Thus to identify the causal real system (with impulse response $h(t) = 0$ for $t < 0$) described by the Laplace transformation $H(s)$ is stable or not, ROC should include jw axis, This will happen if all the poles should lie in the left of the s-plane.

1.7.2 *Inverse Laplace Transformation*

The inverse laplace transformation is computed as $x(t) = \frac{1}{2\pi j}\int_{c-j\infty}^{c+j\infty} X(s)e^{st}ds$ for $t \geq 0$. $x(t)$ is either $\pm\infty$, *notdefined* or *finitevalue* by properly choosing the value for c. If $c = Re(s)$ is chosen in the ROC, $x(t)$ is finite for $t \geq 0$. Consider the computation of Laplace transformation of the signal defined as $\delta(t) + e^{-3t}$ for $t \geq 0$. The laplace transformation is computed as follows:

$$X(s) = \int_{0-}^{\infty} (\delta(t) + e^{-3t})e^{-st}dt$$

$$1 + (\frac{e^{-(s+3)t}}{-(s+3)})_{\infty} - (\frac{e^{-(s+3)t}}{-(s+3)})_{0-}$$

$$1 + (\frac{e^{-(\sigma+j\omega+3)t}}{(-(s+3))})_{\infty} - (\frac{e^{-(\sigma+j\omega+3)t}}{-(s+3)})_{0-}$$

If $(\sigma + 3)$ is chosen greater than zero, then $(\frac{e^{-(\sigma+j\omega+3)t}}{-(s+3)})_{\infty} = 0$ and $(\frac{e^{-(\sigma+j\omega+3)t}}{-(s-3)})_{\infty} = \frac{1}{-(s+3)}$ Thus $X(s)$ is computed as $1 + 0 - (\frac{-1}{s+3}) = 1 + \frac{1}{s+3}$, provided $(\sigma + 3) > 0$. This is the region of convergence. Thus inverse laplace transformation of $1 + \frac{1}{s+3}$ is given as the following.

$$\frac{1}{2\pi j}\int_{c-j\infty}^{c+j\infty} (1 + \frac{1}{s+3})e^{st}ds = \delta(t) + e^{-3t} \tag{1.35}$$

for $t \geq 0$ provided c is chosen as greater than -3.

1.8 Geometrical Interpretation of Computation of Magnitude Response of the Filter Using Pole-Zero Plot of s-Transformation

The magnitude and the phase response of the given transfer function $H(s)$ at $s = jw$ using geometrical measurement is computed as follows:

1. Let the zeros be represented as $z_1, z_2 \cdots z_m$ and the poles be represented as $p_1 p_2 p_3 \cdots p_m$.
2. Compute the euclidean distance between $(0, w)$ and the point described by z_1. Let it be dz_1. This is repeated to obtain $dz_2, dz_3, dz_4, \cdots dz_m$
3. Similarly the euclidean distance between $(0, w)$ and the point described by p_1. Let it be dp_1. This is repeated to obtain $dp_2, dp_3, dp_4, \cdots dz_m$ are computed.
4. Magnitude response of the transfer function at $(0, w)$ is computed as $\frac{\pi_{k=1}^{k=m} dz_k}{\pi_{l=1}^{l=n} dp_l}$
5. Consider the line joining the point $(0, w)$ and the point described by zero z_i makes an angle θ_{zi} with the real line of the pole-zero plot. Let it be. Similarly line joining the point $(0, w)$ and the point described by pole p_i makes an angle θ_{pi} with the real line of the pole-zero plot
6. Similar to magnitude response, the phase response of the transfer function at $(0, w)$ is computed as $\sum_{k=1}^{k=m} \theta_{zk} - \sum_{k=1}^{k=n} \theta_{pk}$

The geometrical method of computing the magnitude response of the transfer function of the form $H(z) = \frac{(z-a)}{z}$ is illustrated in Fig. 1.20.

1.9 Discrete System

- Impulse response of the system: The response of the system to the impulse signal as the input is the impulse response $h(n)$ of the system. The impulse signal $\delta(n)$for the discrete case is as described below.

$$\delta(n) = 1, n = 0$$
$$\delta(n) = 0, \forall\ n \neq 0$$

- Response to the eigensignal to the linear system: If the signal $e^{jw_0 n}$ is the input to the linear system with impulse response $h(n)$ (with transfer function $H(e^{jw_d})$), the output of the system is given as $e^{jw_0 n} H(jw_0)$. The generalized input signal $x(n)$ is represented using Inverse DTFT as $\int_0^{2\pi} X(e^{jw_d}) e^{jw_d} dw_d$ and hence the corresponding output signal $y(n)$ is given as $y(n) = \int_0^{2\pi} X(e^{jw_d}) H(e^{jw_d}) e^{j2\pi ft} dw_d$. Hence $Y(e^{jw_d}) = X(e^{jw_d}) H(e^{jw_d})$. By taking IFT on both sides, we obtain, $y(n) = \sum_{k=-\infty}^{\infty} h(k)x(n-k)$. This is defined as the convolution of the sequence $x(n)$ and $h(n)$

- Linearity: Let $x_i(n)$ be the input to the system and the corresponding output of the system is $y_i(n)$. The system is said to be linear if the output of the system is $\sum_{i=1}^{N} a_i y_i(n)$ corresponding to $\sum_{i=1}^{N} a_i x_i(n)$ as the input to the system.
- Causality: Two arbitrary inputs $x_1(n)$ and $x_2(n)$ are identical, i.e., $x_1(n) = x_2(n)$ for $t \leq n_0$, then the corresponding outputs satisfy $y_1(n) = y_2(n)$ for $n \leq n_0$ for all x_1, x_2 and n_0. The necessary and sufficient condition for system to be causal is $h(n) = 0$ for $n < 0$.
Proof: Consider the impulse response of the system be $h(t)$. Let $x_1(n)$ and $x_2(n)$ are identical for $n \leq n_0$. For $n \leq n_0$

$$y_1(n_0) = \sum_{k=-\infty}^{n_0} h(k)x_1(n-k)$$

$$\Rightarrow y_1(n) = \sum_{k=-\infty}^{n_0} h(k)x_1(n-k)$$

$$y_2(n_0) = \sum_{k=-\infty}^{n_0} h(k)x_2(n-k)$$

$$\Rightarrow y_2(n_0) = \sum_{k=0}^{n_0} h(k)x_2(n-k)$$

$$\Rightarrow y_1(n_0) = y_2(n_0)$$

Thus if $h(k) = 0$ for $k \leq 0$, the system is the causal system. Consider the system with $h(-5) \neq 0$ and let $x_1(n) = 0 \; \forall \, n$ and $x_2(n) = \delta(n-5)$. It is observed that the values of $x_1(n)$ and $x_2(n)$ are identical upto $n \leq 0$, the values for $y_1(0)$ and $y_2(0)$ are computed as follows.

$$y_1(0) = \sum_{k=-\infty}^{0} h(k)x_1(-k) = 0$$

$$y_2(0) = \sum_{k=-\infty}^{0} h(k)x_2(-k) = x_2(5)$$

$y_1(0) \neq y_2(0)$ and hence $h(k)$ is causal only if $h(k) = 0$ for $k \leq 0$. Hence proved.
- Stability: A signal is bounded if its magnitude is limited to a finite value at all times. If the bounded input gives bounded output (BIBO), it is the stable system. The system with impulse response $h(n)$ is BIBO stable if and only

$$\sum_{k=-\infty}^{k=\infty} |h(k)|d \leq \infty \tag{1.36}$$

Proof: Using Schwartz inequality,

$$y(k) = \sum_{k=-\infty}^{k=\infty} h(k)x(n-k)$$

$$|y(k)| \leq \sum_{k=-\infty}^{k=\infty} |h(k)| \sum_{k=-\infty}^{k=\infty} |x(n-k)|$$
$$\leq K \int_{-\infty}^{\infty} |h(k)|$$

Thus if $\sum_{k=-\infty}^{k=\infty} |h(k)|d \leq \infty$, $|y(k)| \leq \infty$, i.e., stable system.
Suppose that $h(k)$ is not satisfying $\sum_{k=-\infty}^{k=\infty} |h(k)|d \leq \infty$ and let for instance, there exists some finite n_0 such that $\sum_{k=-\infty}^{n_0} |h(k)| = \infty$. Consider the case the input satisfy $x(n-n_0) = -1$ for $h(n) \geq 0$ and $x(n-n_0) = -1$ for $h(n) \leq 0$. This implies the following.

$$y(n_0) = \sum_{k=-\infty}^{k=n_0} h(k)x(n-k) = \sum_{k=-\infty}^{k=n_0} |h(k)| = \infty$$

Thus the system is stable only if $\sum_{-\infty}^{\infty} |h(k)| \leq \infty$. Hence proved.

- Time in-variant system: Let $y(n)$ is the output corresponds to the signal $x(n)$, then the system is said to time in-variant if $y(n-n_0)$ is the output corresponds to the input $x(n-n_0)$
- Memory system: If the output of the system $y(n)$ depends on $x(n+n_0)$ and/or $y(n+n_0)$, with $n_0 = 0$, then the system is said to memory-less system.
- Linear phase of the system: Let the transfer function of the system is represented as $|H(e^{jw_d})|e^{j\phi(w_d)}$. The magnitude response is given as $|H(e^{jw_d})|$ and the phase response is given as $\phi(w_d)$, Let the output signal corresponding to the input signal $x(n) = a_1e^{jw_{d_1}n} + a_2e^{jw_{d_2}n}$ is obtained as $a_1e^{jw_{d_1}n}|H(e^{jw_{d_1}})|e^{j\phi(w_{d_1})} + a_2e^{jw_{d_2}n}|H(e^{jw_{d_2}})|e^{j\phi(w_{d_2})}$ respectively. Suppose the input signal is delayed by n_0,the corresponding output is given as the following.

$$a_1e^{jw_{d_1}(n-n_0)}|H(e^{jw_{d_1}})|e^{j\phi(w_{d_1})} + a_2e^{jw_{d_2}(n-n_0)}|H(e^{jw_{d_2}})|e^{j\phi(w_{d_2})}$$

which may not be equal to $y(n-n_0)$. But suppose if we choose $\phi(w_d) = -\tau w_d$, where τ is the integer constant, we get the output as the following.

$$a_1e^{jw_{d_1}(n-n_0)}|H(e^{jw_{d_1}})|e^{j\phi(w_{d_1})} + a_2e^{jw_{d_2}(n-n_0)}|H(e^{jw_{d_2}})|e^{j\phi(w_{d_2})}$$
$$= a_1e^{jw_{d_1}(n-n_0-\tau)}|H(e^{jw_{d_1}})| + a_2e^{j2\pi f_2(n-n_0-\tau)}|H(e^{jw_{d_2}})|$$
$$= y(n-n_0)$$

Thus if the phase response is linear, we get undistorted delayed version of the output signal if the linear combinations of the delayed eigensignals are given as the input to the linear system.

1.10 Z-Transformation

The z-transformation of the right-sided sequence is given as the following.

$$X(z) = \sum_{k=0}^{k=\infty} x(k)z^{-k} \tag{1.37}$$

where $z = re^{iw_d}$. The z-plane is the complex plane with each point represents the complex number represented in the polar form. The set of all values of z for which $X(z)$ (Computed for the particular sequence) is finite is the region of convergence (ROC). Let us consider computing the z-transformation for $a^n u(n)$, where $u(n)$ is the unit step function as shown below.

$$X(z) = \sum_{k=0}^{k=\infty} a^k z^{-k} = 1 + (az^{-1}) + (az^{-1})^2 + \cdots \tag{1.38}$$

The above equation converges if $|az^{-1}| < 1$ and the corresponding z-transformation is given as $(1 - az^{-1})^{-1} = \frac{z}{z-a}$. The region $|z| > |a|$ is known as region of convergence (ROC) (refer Fig. 1.19). It is observed that if $r = 1$, z-transformation becomes Discrete Time Fourier transformation. Hence if $X(z)$ is finite for all values of w_d with $r = 1$, we conclude that Discrete Fourier transformation exists. In other words, Discrete Time Fourier transformation exists if the ROC includes the $r = 1$ circle. Thus Discrete Time Fourier tansformation can be viewed as the special case of z-transformation with $r = 1$. The set of all values of z for which $X(z) = 0$ are the zeros of the z-transformation $X(z)$. Similarly, the set of all values of z for which $X(z) = \infty$ are the poles of the $X(z)$.

1.10.1 Properties of the ROC of the Z-Transformation for the Right-Sided Sequence

1. ROC is of the circular strip shape in the z-plane.
2. For the Right Sided Sequence (R.S.S), ROC is in the outside the circle passes through the poles of the z-plane.

3. Hence to include $r = 1$ circle in the ROC, all the poles of the z-transformation should lie inside the unit circle. In this context, the transfer function $H(z)$ is said to be stable if all the poles lie inside the unit circle.
4. ROC does not contain the poles of $H(z)$.
5. Thus to identify the causal real system (with impulse response $h(n) = 0$ for $n < 0$) described by the z-transformation $H(z)$ is stable or not, ROC should include jw axis, this will happen if all the poles should lie in the left of the z-plane.

1.10.2 *Inverse Z-Transformation*

The inverse z-transformation of the $X(z)$ is obtained as the contour integration as given below.

$$\frac{1}{2\pi j}\oint X(z)z^{k-1}dz \tag{1.39}$$

for $k \geq 0$. With the contour selected as the circle with 0 as the origin and lies in the region of convergence. Consider the z-transformation of the sequence $x(n) = 0.5^n$ for $n \geq 0$. The z-transformation is obtained as the following.

$$\begin{aligned} X(z) &= \sum_{k=0}^{k=\infty} x(k)z^{-k} \\ &= \sum_{k=0}^{k=\infty} 0.5^k z^{-k} \\ &= \sum_{k=0}^{k=\infty} (0.5z^{-1})^k \\ &= 1 + 0.5z^{-1} + (0.5z^{-1})^2 + \cdots \end{aligned}$$

If $|0.5z^{-1}| < 1$,i.e., $|z| = |re^{jw_d}| = |r| > 0.5$ we get the following. $1 + 0.5z^{-1} + (0.5z^{-1})^2 + \cdots = (1 - 0.5z^{-1})^{-1} = \frac{z}{z-0.5}$. Thus the inverse z-transformation of $\frac{z}{z-0.5}$ is computed as the following.

$$\frac{1}{2\pi j}\oint \frac{z}{z-0.5}z^{k-1}dz = 0.5^n \tag{1.40}$$

for $n \geq 0$, provided the contour is chosen as the circle with the radius greater than 0.5.

1.11 Response of the Digital Filter with the Typical Transfer Function (Obtained Using DTFT) to the Periodic Sequence

Let the magnitude response of the digital filter be represented as $H(e^{jw_d})$. The response of the filter to the periodic sequence $x(n)$ is obtained as follows.

$$x(n) = \frac{1}{N}\sum_{k=0}^{k=N-1} X(k)e^{\frac{j2\pi nk}{N}} \tag{1.41}$$

The response to the eigensignal $e^{\frac{j2\pi nk}{N}}$ is obtained as $e^{\frac{j2\pi nk}{N}}\ H(e^{\frac{j2\pi k}{N}})$. As the system is linear, the response to the periodic sequence $x(n)$ is obtained as the following.

$$y(n) = \frac{1}{N}\sum_{k=0}^{k=N-1} X(k)e^{\frac{j2\pi nk}{N}} H(e^{\frac{j2\pi k}{N}}) \tag{1.42}$$

This is demonstrated using the filter coefficients h (refer Fig. 1.22b) and the corresponding transfer function is given in the Fig. 1.21. The response of the system described by the impulse response h to the input periodic signal x (Fig. 1.22a) is obtained as the convolution of periodic sequence x and the filer coefficients h as shown in Fig. 1.22c. Let $X(k)$ be the DFT sequence corresponding to the input sequence $x(n)$. Then the nth value of the output sequence y is also obtained by computing the linear combinations of the samples given in the Fig. 1.21 ($H(e^{\frac{j2\pi k}{N}})$) and the sequence $X(k)e^{\frac{j2\pi nk}{N}}$ (computed for the typical value of n) as shown in Fig. 1.22d. It is observed that the output obtained in the Fig. 1.22 c, d are identical.

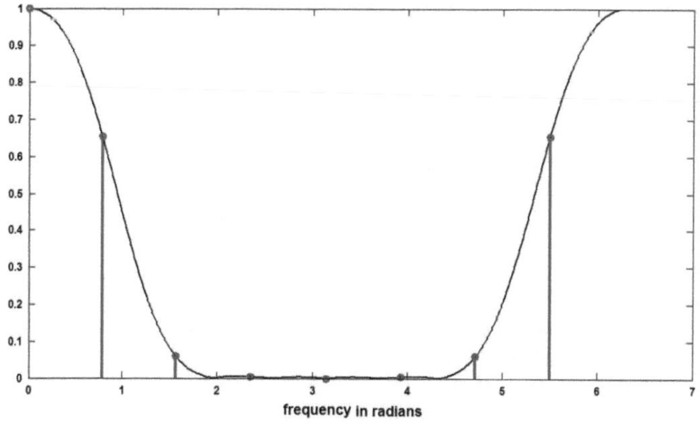

Fig. 1.21 Magnitude response of the system used for the demonstration. The samples used to obtain the response of the periodic sequence are indicated in *red color*

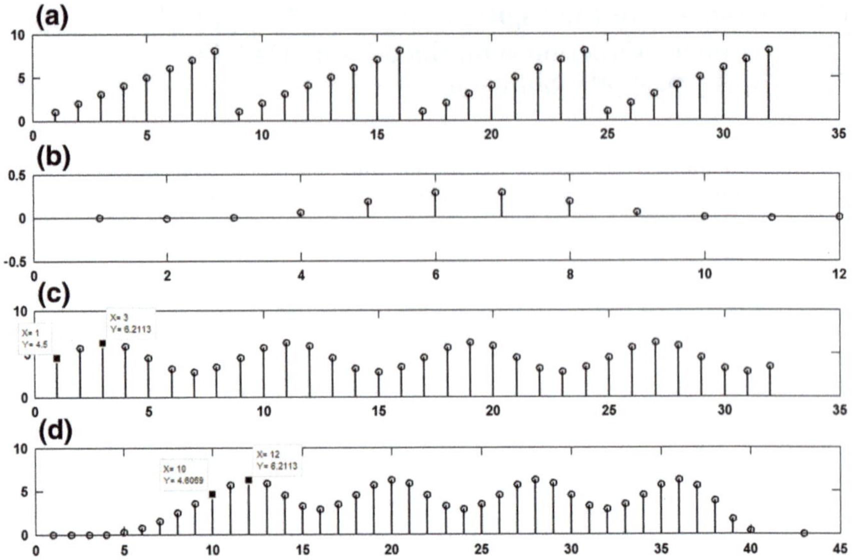

Fig. 1.22 **a** Periodic input sequence. **b** Impulse response of the filter. **c** Output sequence computed using convolution. **d** Output sequence computed using linear combinations of samples mentioned in the Fig. 1.21

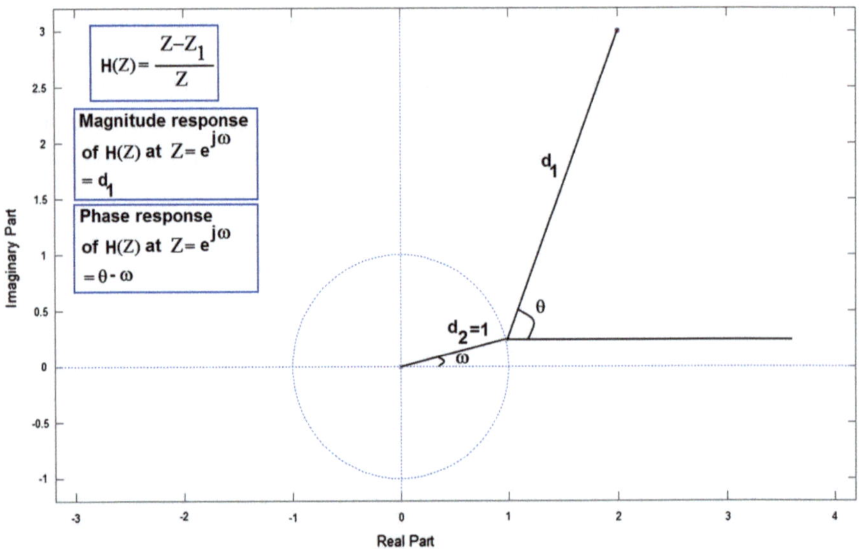

Fig. 1.23 Illustration of the Geometrical method of computing magnitude and phase response of the discrete system transfer function $H(z)$

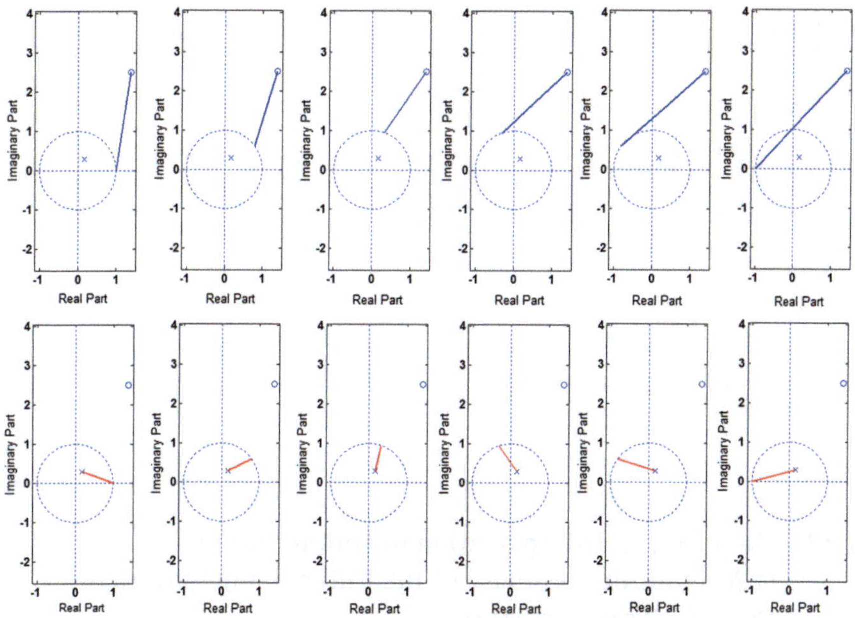

Fig. 1.24 Illustration of the Geometrical method of computing magnitude and phase response of the discrete system transfer function H(z) (contd.)

```
%responsetoperiodicsequence.m
%Response to the periodic sequence
h=fir1(11,0.3);
t=0:1:511;
POINTS=512;
N=8;
[H,W]=freqz(h,1,2*pi*t/POINTS);
input=1:1:8;
y=fft(input);
output1=[];
for n=1:1:32
temp=0;
for i=0:1:7
temp=temp+ y(i+1)*exp(j*2*pi*n*i/N)*HRES(2*pi*i/N,h);
end
output1=[output1 temp/(N)];
end
%Considering four period
I=[input input input input]
output2=conv(h,I)
figure
subplot(4,1,1)
stem(I)
subplot(4,1,2)
stem(h)
subplot(4,1,3)
stem(real(output1))
subplot(4,1,4)
```

```
stem(output2,'r')
k=1;
temp1=[];
temp2=[];
for i=0:1:7
temp1=[temp1 2*pi*i/N];
temp2=[temp2 HRES(2*pi*i/N,h)];
end
SAMPLE=temp1;
VALUE=temp2;
figure
plot(W,abs(H));
hold on
stem(SAMPLE,abs(VALUE),'r')

function [res]=HRES(w,h)
n=0:1:length(h)-1;
res=sum(h.*exp(-j*w*n));
```

1.12 Geometrical Interpretation of Computation of Magnitude Response of the Filter Using Pole-Zero Plot of z-Transformation

The magnitude and the phase response of the given transfer function $H(z)$ at $z = e^{jw}$ using geometrical measurement is computed as follows:

1. Let the zeros be represented as $z_1, z_2 \cdots z_m$ and the poles be represented as $p_1 p_2 p_3 \cdots p_m$. If the system is all zero system, pole at orgin is considered. Similarly if the system is all pole system, zero at orgin is considered
2. Compute the euclidean distance between $(cos(w), sin(w))$ and the point described by z_1. Let it be dz_1. This is repeated to obtain $dz_2, dz_3, dz_4, \cdots dz_m$
3. Similarly the euclidean distance between $(cos(w), sin(w))$ and the point described by p_1. Let it be dp_1. This is repeated to obtain $dp_2, dp_3, dp_4, \cdots dz_m$ are computed.
4. Magnitude response of the transfer function at $(cos(w), sin(w))$ is computed as $\frac{\pi_{k=1}^{k=m} dz_k}{\pi_{l=1}^{l=n} dp_l}$
5. Consider the line joining the point $(cos(w), sin(w))$ and the point described by zero z_i makes an angle θ_{zi} with the real line of the pole-zero plot. Let it be. Similarly line joining the point $(cos(w), sin(w))$ and the point described by pole p_i makes an angle θ_{pi} with the real line of the pole-zero plot
6. Similar to magnitude response, the phase response of the transfer function at $(cos(w), sin(w))$ is computed as $\sum_{k=1}^{k=m} \theta_{zk} - \sum_{k=1}^{k=n} \theta_{pk}$

The geometrical method of computing the magnitude response and phase response of the transfer function of the form $H(z) = \frac{(z-a)}{z}$ is illustrated in Figs. 1.23, 1.24 and 1.25.

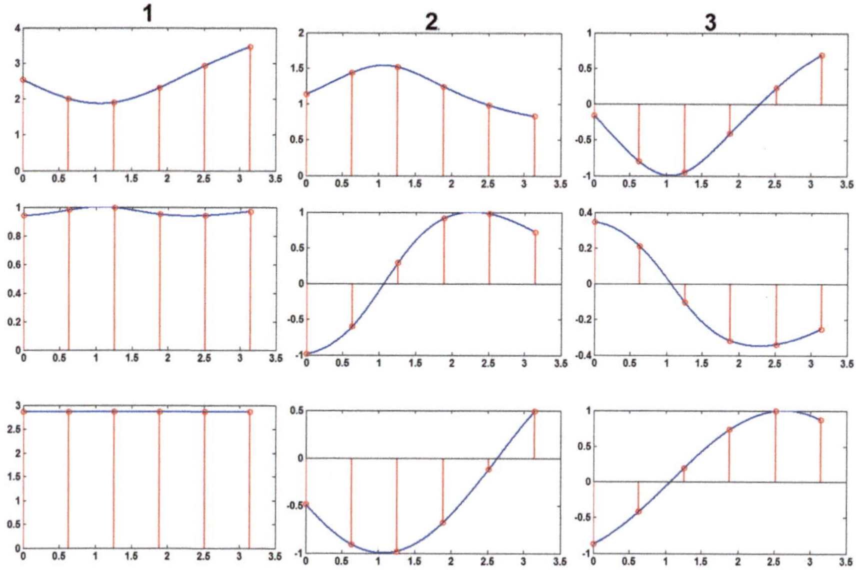

Fig. 1.25 Column 1 indicate the actual magnitude response of continuous curve in the typical filter. Column 2 indicate the cos of the phase response of the filter. Column 3 indicate the sine of the phase response of the filter. The stem in all the subplots indicate the corresponding values computed using the Geometrical method

```
%complexplanegeometry.m
%Demonstration of the geometrical method of computing magnitude
and phase response
z1=1.4+ 2.5*j;
z2=1/z1';
[mag1,mag2,phase1,phase2]=magrespzplot1(z1,z2);
[H1,W1]=freqz([1 -z1],1);
[H2,W2]=freqz(1,[1 -z2]);
w=linspace(0,pi,6);
figure
subplot(3,3,1)
plot(W1, abs(H1))
hold on
stem(w,mag1,'r')
subplot(3,3,2)
plot(W2, abs(H2))
hold on
stem(w,mag2,'r')
subplot(3,3,3)
plot(W1, cos(angle(H1)))
hold on
stem(w,cos(phase1),'r')
subplot(3,3,4)
plot(W2, cos(angle(H2)))
hold on
stem(w,cos(phase2),'r')
```

```
subplot(3,3,5)
plot(W1, sin(angle(H1)))
hold on
stem(w,sin((phase1)),'r')
subplot(3,3,6)
plot(W2, sin(angle(H2)))
hold on
stem(w,sin(phase2),'r')
subplot(3,3,7)
plot(W1, abs(H1).*abs(H2))
hold on
stem(w,mag1.*mag2,'r')
subplot(3,3,8)
plot(W2, cos(angle(H1)-angle(H2)))
hold on
stem(w,cos(phase1-phase2),'r')
subplot(3,3,9)
plot(W2, sin(angle(H1)-angle(H2)))
hold on
stem(w,sin(phase1-phase2),'r')


function [MAG1,MAG2,PHASE1,PHASE2]=magrespzplot(z,p)
%The zero at orgin is considered when the phase response is
%computed for pole at p
%The pole at orgin is considered when the phase response is
%computed for zero at z
MAG1=[];
MAG2=[];
PHASE1=[];
PHASE2=[]
f=1;
for w=linspace(0,pi,6)
figure(1)
subplot(2,6,f)
zplane(z,p)
hold on
subplot(2,6,f+6)
zplane(z,p)
hold on
w1=exp(j*w);
figure(1)
subplot(2,6,f)
[real(z) imag(z)];
plot([real(w1) real(z)],[imag(w1) imag(z)],'b-');
hold on
figure(1)
subplot(2,6,f+6)
[real(w1),imag(w1)];
[real(p) imag(p)];
plot([real(w1),real(p)],[imag(w1) imag(p)],'r-');
hold on
MAG1=[MAG1 sqrt(abs((w1-z).^2))];
MAG2=[MAG2 1./sqrt(abs((w1-p).^2))];
PHASE1=[PHASE1 angle(w1-z)-angle(w1)];
PHASE2=[PHASE2 angle(w1)-angle(w1-p)];
f=f+1;
end
```

Chapter 2
Infinite Impulse Response (IIR) Filter

2.1 Impulse-Invariant Mapping

The generalized transfer function of the system can be represented in Laplace transformation as given below:

$$H_a(s) = \sum_{k=1}^{k=N} \frac{A_k}{(S - p_k)}. \tag{2.1}$$

The corresponding impulse response of the causal system is obtained as

$$h_a(t) = \sum_{k=1}^{k=N} A_k e^{-jwt} u(t), \tag{2.2}$$

where $u(t)$ is the unit step function. Sampling the impulse response $h_a(t) = h(t)$, we get the discrete version of the system as given below:

$$h_a(nT_s) = h(n) = \sum_{k=1}^{k=N} A_k e^{-jwnT_s}, \tag{2.3}$$

for $n = 0\ 1 \ldots$. Taking z-transformation of the sequence $h(n)$, we get the following (Fig. 2.1):

$$H(z) = \sum_{n=0}^{n=\infty} h(n) z^{-n}$$

$$= \sum_{n=0}^{n=\infty} \sum_{k=1}^{k=N} A_k e^{-jwnT_s}$$

E.S. Gopi, *Multi-Disciplinary Digital Signal Processing*,
DOI 10.1007/978-3-319-57430-1_2

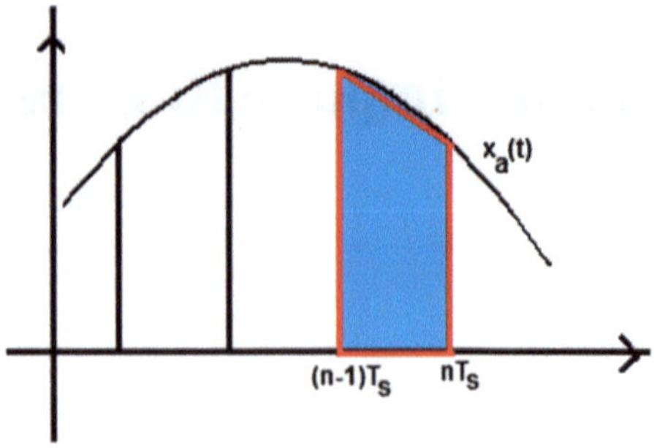

Fig. 2.1 Illustration on the computation of the area under the curve using Trapezoidal rule

$$= \sum_{k=1}^{k=N} A_k \sum_{n=0}^{n=\infty} e^{-jwnT_s} z^{-n}$$

$$= \sum_{k=1}^{k=N} A_k \sum_{n=0}^{n=\infty} (e^{-jwT_s} z^{-1})^n$$

$$= \sum_{k=1}^{k=N} A_k (1 - e^{-jwT_s} z^{-1})^{-1}$$

$$= \sum_{k=1}^{k=N} \frac{A_k}{1 - e^{-jwT_s} z^{-1}}.$$

By substituting $\frac{A_k}{(S-p_k)}$ with $\frac{A_k}{1-e^{-jwT_s} z^{-1}}$, we convert the continuous domain to discrete domain, i.e., we obtain the discrete sequence $h(n)$ from the continuous impulse response $h(t)$. This the impulse-invariant method of mapping S to Z domain.

2.2 Bilinear Transformation Mapping

The sampling frequency $F_s = \frac{1}{T_s}$ used in (2.3) needs to be fixed as greater than twice the maximum frequency content of $h(t)$ (Sampling theorem), which is not usually known. Suppose if F_s is chosen not satisfying the sampling theorem associated with the impulse response, overlapping in the spectrum occurs. In particular, if the spectrum of $h(t)$ is high-pass nature, it suffers a lot. This is circumvented using the technique known as bilinear transformation as described below. The area under the curve of the impulse response $x_a(t) = \frac{dh(t)}{dt}$ for $(n-1)T_s \leq t \leq nT_s$ is computed (refer Fig. 2.1) as

$$\int_{(n-t)T_s}^{nT_s} \frac{dh(t)}{dt} dt = (h((n-1)T_s) - h(nT_s)). \tag{2.4}$$

This is computed using the trapezoidal approximation (refer Fig. 2.1) as follows:

$$\frac{T_s}{2}(x_{nT_s} + x_{(n-1)T_s}). \tag{2.5}$$

Taking Laplace transformation on both sides of $x(t) = \frac{dh(t)}{dt}$, we get $X(s) = sH(s)$. Taking z-transformation of (2.4) and (2.5), we get the following: $H(z)(1 - z^{-1}) = \frac{T_s}{2}X(z)(1 + z^{-1})$. Thus equating the ratio $\frac{X(s)}{H(s)}$ with $\frac{X(z)}{H(z)}$, we get the following:

$$s = \frac{2}{T_s}\frac{1 - z^{-1}}{1 + z^{-1}}. \tag{2.6}$$

Substituting $s = jw$ and $z = e^{jw_d}$ in (2.6), we get $w = \frac{2}{T_s}tan(w_d/2)$, where w is the analog frequency and w_d is the digital frequency. This method of mapping s-domain to z-domain is called Bilinear transformation. Even when w tends to ∞, w_d tends to the value π. As π corresponds to the maximum frequency content of the signal after sampling, maximum frequency of the content of the signal is bounded to ∞. This is equivalent to obtaining the scaled down version of the spectrum of $h(t)$ such that maximum frequency is bounded to $\frac{F_s}{2}$, irrespective of actual value of the F_s. Thus overlapping of spectrum never occur. Hence, this is suitable for high-pass filtering. But the drawback is the shrinkage of the spectrum.

2.2.1 *Frequency Pre-warping*

The relationship between the digital frequency w_d and the analog frequency w (rad/sec) is linear ($w_d = wT_s$) in the case of impulse-invariant mapping. But if T_s is not properly chosen to obtain the discrete version of the analog filter, overlapping occurs. This is circumvented using Bilinear transformation given as $w = \frac{2}{Ts}tan(\frac{w_d}{2})$. This guarantees that even when the maximum analog frequency content of the impulse response is ∞, the corresponding digital frequency is bounded to π. But the relationship is nonlinear (refer Fig 2.2). Suppose if would like to design the low-pass filter with cutoff frequency w_c in rad/sec (equivalently $\frac{w_c}{F_s}$ in digital domain), we get the digital filter with cutoff frequency $2\frac{wT_s}{2}tan^{-1}(w)$. This is undesired property of Bilinear transformation. This is circumvented as follows.

- Suppose we need the low-pass filter with cutoff frequency w_c in rad/sec (equivalently $\frac{w_c}{F_s}$ in digital domain), we obtain the pre-warped frequency $pw_c = \frac{2}{Ts}tan$ $(\frac{w_cT_s}{2})$. The plot between w_c and pw_c is given in the bandpass filter as shown in Fig. 2.3.

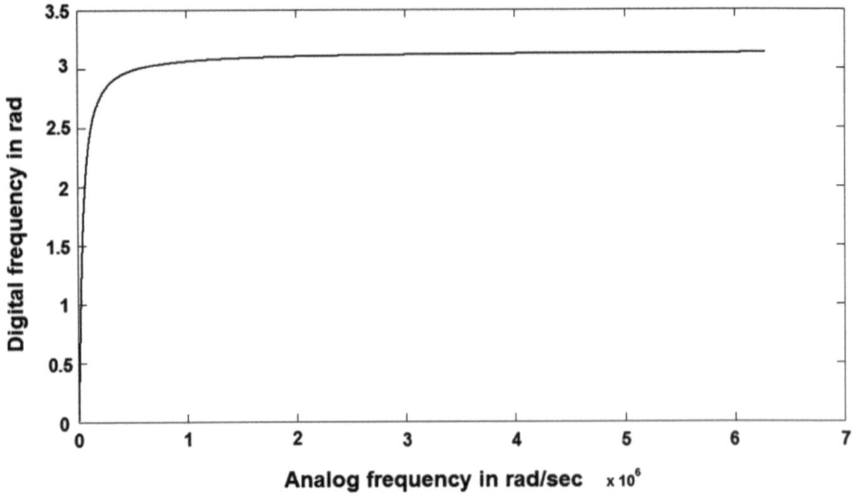

Fig. 2.2 Relationship between digital and analog frequency using bilinear transformation

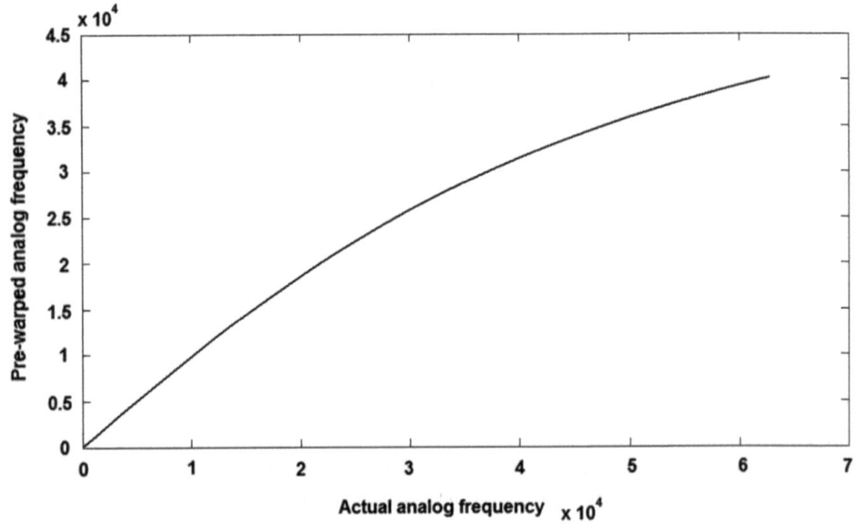

Fig. 2.3 Relationship between the actual analog frequency and the prewarped analog frequency (refer (2.2.1))

- Design the analog filter with the prewarped frequency pw_c in rad/sec.
- If the mapping is done from s to z using bilinear transformation for the designed analog filter, we get the digital filter with the desired cutoff frequency $\frac{w_c}{F_s}$.
- This is known as frequency pre-warping.

2.2.2 Design of Digital IIR Filter using Butterworth Analog Filter and Impulse-Invariant Transformation

- The generalized transfer function of the IIR analog low-pass filter is computed as follows:

$$H_a(s) = \pi_{k=1}^{k=\frac{N}{2}} \frac{B_k w_c^2}{S^2 + b_k w_c s + c_k w_c^2} \tag{2.7}$$

for N as even.

$$H_a(s) = \pi_{k=1}^{k=\frac{N-1}{2}} \frac{B_k w_c^2}{S^2 + b_k w_c s + c_k w_c^2} \frac{B_0 w_c}{s + c_0 w_c} \tag{2.8}$$

for N as odd.
- The magnitude response of the Butterworth filter is given as follows:

$$|H(jw)| = \frac{A}{[1 + (\frac{w}{w_c})^{2N}]^{\frac{1}{2}}}. \tag{2.9}$$

Refer Fig.2.4 for the typical magnitude response plot for various orders of the Butterworth filter with $w_c = 1$ rad/sec and $A = 1$.
- Given the magnitude response of the Butterworth filter at $w = 0$ (say A), magnitude of the transfer function is lesser than m at the stop band frequency (ws in rad/sec), and the order of the filter N is computed as follows: $N = \frac{\frac{A^2}{m^2} - 1}{2log(\frac{ws}{wc})}$, where wc is the cutoff of the Butterworth filter whose magnitude response is $\frac{A}{\sqrt{2}}$ at wc.
- For the typical value of N as even, the values for b_k are computed as $b_k = sin(\frac{(2k-1)\pi}{2N})$, $c_k = 1$, $B_k = A^{\frac{2}{N}}$ for $k = 1 \cdots \frac{N}{2}$.
- For the typical value of N as odd, the values for b_k are computed as $b_k = sin(\frac{(2k-1)\pi}{2N})$, $c_k = 1$, $B_k = A^{\frac{2}{N}}$ for $k = 1 \cdots \frac{N-1}{2}$ and $B_0 = 1$ and $c_0 = 1$.
- Mapping from the s-domain to z-domain ($H(S)$ to $H(z)$) is obtained by substituting the term of the form $\frac{A_k}{(S-p_k)}$ of $H_a(s)$ with $\frac{A_k}{1-e^{-jwT_s}z^{-1}}$. This is done by representing $\frac{B_k w_c^2}{S^2+b_k w_c s+c_k w_c^2}$ as the summation of two partial fractions for every k.
- Thus the digital Butterworth impulse-invariant filter $H(z)$ is obtained.

2.2.3 Design of Digital IIR Filter using Butterworth Analog Filter and Bilinear Transformation

- The magnitude response of the Butterworth filter is given as follows (2.9):

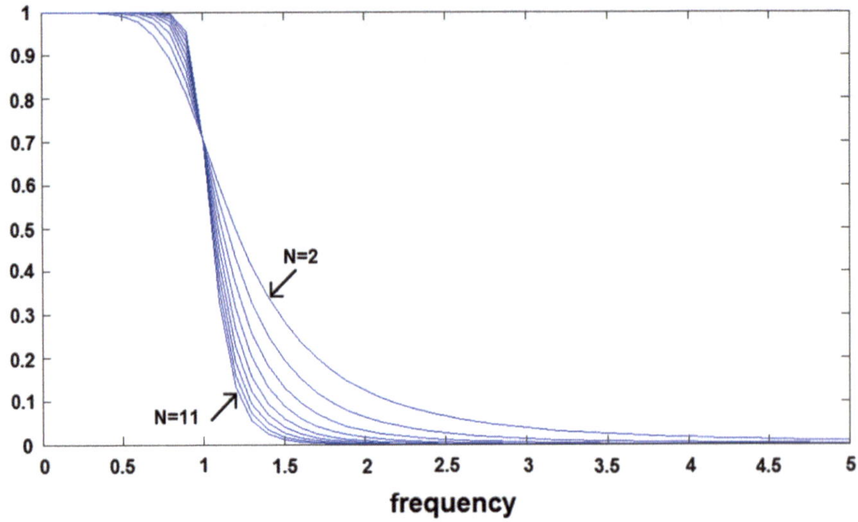

Fig. 2.4 Magnitude response plot for various orders of the Butterworth filter with $w_c = 1$ rad/sec and $A = 1$

- Given the magnitude response of the Butterworth filter at $w = 0$ (say A), magnitude of the transfer function is lesser than m at the stop band frequency (ws in rad/sec), cutoff frequency (wc in rad/sec), and sampling frequency Fs, and the order of the filter N is computed as follows.
- Obtain the prewarped frequency corresponding to ws and wc as pws and pwc as follows:

$$wcd = \frac{w_c}{Fs};\ wsd = \frac{w_s}{Fs};$$

$$pwc = \frac{2}{Ts}tan(\frac{wcd}{2});\ pws = \frac{2}{Ts}tan(\frac{wsd}{2});$$

- The order of the filter N is computed as $N = \frac{\frac{A^2}{m^2}-1}{2log(\frac{pws}{pwc})}$.
- For the typical value of N as even, the values for b_k are computed as $b_k = sin(\frac{(2k-1)\pi}{2N})$, $c_k = 1$, $B_k = A^{\frac{2}{N}}$ for $k = 1 \cdots \frac{N}{2}$.
- For the typical value of N as odd, the values for b_k are computed as $b_k = sin(\frac{(2k-1)\pi}{2N})$, $c_k = 1$, $B_k = A^{\frac{2}{N}}$ for $k = 1 \cdots \frac{N-1}{2}$ and $B_0 = 1$ and $c_0 = 1$. Thus the analog filter $H_a(s)$ is obtained (refer (1.7) and (1.8)).
- Mapping from the s-domain to z-domain ($H(S)$ to $H(z)$) is obtained by substituting $s = \frac{2}{Ts}\frac{1-z^{-1}}{1+z^{-1}}$.
- Thus the bilinear transformation-based Butterworth filter $H(z)$ is obtained.

2.2.4 Design of Digital IIR Filter Using Chebyshev Analog Filter and Impulse-Invariant Transformation

- Butterworth filter has the smooth magnitude response, but the cutoff is not usually very sharp. This is circumvented using the chebyfilter analog filter.
- The magnitude response of the Chebyshev filter is given as follows:

$$|H(jw)| = \frac{A}{[1 + \varepsilon^2 C_N(\frac{w}{w_c})]^{\frac{1}{2}}}, \tag{2.10}$$

where $C_N(x) = cos(Ncos^{-1}x)$ for $x \leq 1$ and $C_N(x) = cosh(Ncosh^{-1}x)$ for $x > 1$ (refer Fig. 2.5 for the typical magnitude response plot for various orders (red color for N odd and blue color for N as even) is of the Chebyshev filter with $w_c = 2$ rad/sec, $A = 1$ and $\varepsilon = 0.2$. Also, Fig. 2.6 shows the case when $\varepsilon = 2$.
- Given the ripple width R, maximum amplitude of the transfer function A, pass band cutoff frequency $w_p = w_c$ (magnitude at w_c is given as $\frac{A}{\sqrt{1+\varepsilon^2)}}$), where $(A - \frac{A}{\sqrt{1+\varepsilon^2}})$ is the ripple width R and the magnitude of the transfer function at the stop band frequency (ws in rad/sec) is lesser than m, the order of the filter N is computed as follows:

$$\varepsilon = \sqrt{\frac{R}{A - R}};\ r = \frac{w_s}{w_p};\ C = \sqrt{\frac{1 - m^2}{m^2 \varepsilon^2}};\ N = \frac{acosh(C)}{acosh(r)}.$$

- For the typical value of N, the values for b_k are computed as $b_k = 2Y_N sin(\frac{(2k-1)\pi}{2N})$, $c_k = (Y_N)^2 + cos^2(\frac{(2k-1)\pi}{2N})$, where $Y_N = \frac{1}{2}([\frac{1}{\varepsilon}^2 + \frac{1}{\varepsilon}]^{\frac{1}{N}} + [\frac{1}{\varepsilon}^2 - \frac{1}{\varepsilon}]^{\frac{-1}{N}})$ and B_k is chosen by choosing the required amplitude (either A for $N = odd$ or $\frac{A}{(1+\varepsilon^2)^{\frac{1}{2}}}$ for $N = even$ at $w = 0$).
- Mapping from the s domain to z domain ($H(S)$ to $H(z)$) is obtained by substituting the term of the form $\frac{A_k}{(S-p_k)}$ of $H_a(s)$ with $\frac{A_k}{1-e^{-jwT_s}z^{-1}}$. This is done by representing $\frac{B_k w_c^2}{S^2+b_k w_c s + c_k w_c^2}$ as the summation of two partial fractions for every k.
- Thus the digital Butterworth impulse-invariant filter $H(z)$ is obtained.

2.2.5 Design of Digital IIR Filter Using Chebyshev Analog Filter and Bilinear Transformation

- Butterworth filter has the smooth magnitude response, but the cutoff is not usually very sharp. This is circumvented using the chebyshev analog filter.
- The magnitude response of the Chebyshev filter is given as (2.10).

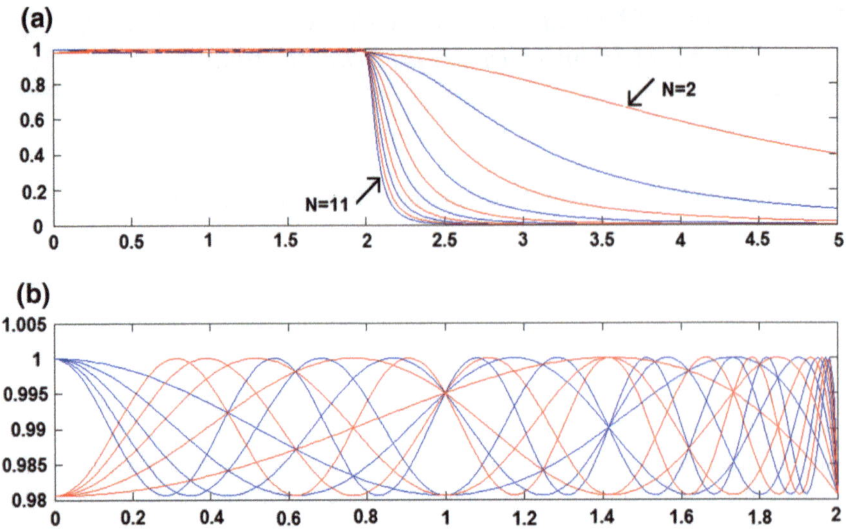

Fig. 2.5 Magnitude response plot for various orders (*red color* for N odd and *blue color* for N as even) is of the Chebyshev filter with $w_c = w_p = 2$ rad/sec (refer Sects. 2.2.4 and 2.2.5), $A = 1$ and $\varepsilon = 0.2$

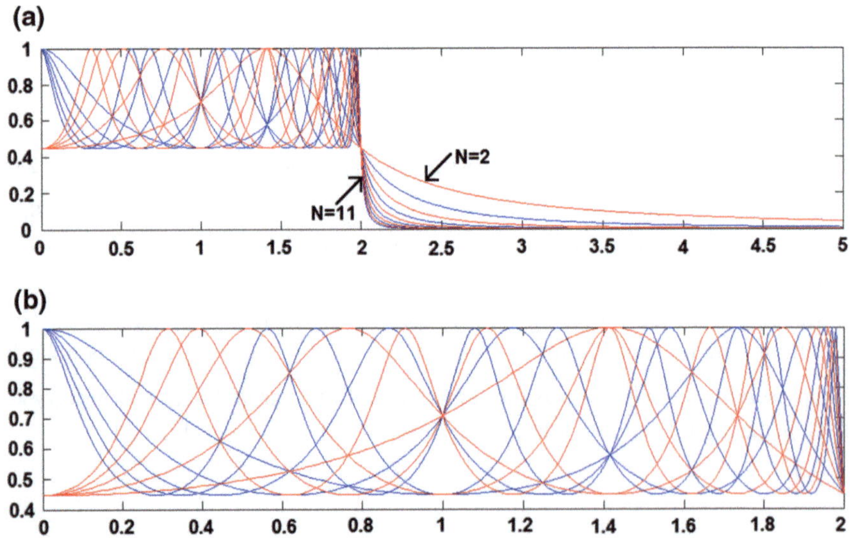

Fig. 2.6 Magnitude response plot for various orders (*red color* for N odd and *blue color* for N as even) is of the Chebyshev filter with $w_c = w_p = 2$ rad/sec, $A = 1$ and $\varepsilon = 2$

- Given the ripple width R, maximum amplitude of the transfer function A, pass band cutoff frequency $w_p = w_c$ (magnitude at w_c is given as $\frac{A}{\sqrt{1+\varepsilon^2)}}$, where $(A - \frac{A}{\sqrt{1+\varepsilon^2}})$ is the ripple width R and the magnitude of the transfer function at the stop band frequency (ws in rad/sec) is lesser than m), the order of the filter N is computed as follows.
- Obtain the prewarped frequency corresponding to ws and wc as pws and pwc as follows:

$$wcd = \frac{w_c}{Fs}; wsd = \frac{w_s}{Fs};$$

$$pwc = \frac{2}{Ts} tan(\frac{wcd}{2}); pws = \frac{2}{Ts} tan(\frac{wsd}{2});$$

$$\varepsilon = \sqrt{\frac{R}{A-R}}; r = \frac{pws}{pwp}; C = \sqrt{\frac{1-m^2}{m^2\varepsilon^2}}; N = \frac{acosh(C)}{acosh(r)}.$$

- For the typical value of N, the values for b_k are computed as $b_k = 2Y_N sin(\frac{(2k-1)\pi}{2N})$, $c_k = (Y_N)^2 + cos^2(\frac{(2k-1)\pi}{2N})$, where $Y_N = \frac{1}{2}([\frac{1}{\varepsilon}^2 + \frac{1}{\varepsilon}]^{\frac{1}{N}} + [\frac{1}{\varepsilon}^2 - \frac{1}{\varepsilon}]^{\frac{-1}{N}})$ and B_k is chosen by choosing the required amplitude (either A for $N = odd$ or $\frac{A}{(1+\varepsilon^2)^{\frac{1}{2}}}$) for $N = even$ at $w = 0$).
- Mapping from the s-domain to z-domain ($H(S)$ to $H(z)$) is obtained by substituting $s = \frac{2}{Ts}\frac{1-z^{-1}}{1+z^{-1}}$.
- Thus the bilinear transformation-based Chebyshev filter $H(z)$ is obtained.

```
%plotbuttermag.m
%Magnitude response of the Butterworth filter
function [res]=plotbuttermag(A,fc)
figure
for N=3:1:11
f=0:0.1:5;
M=A./(1+(f/fc).^(2*N)).^(1/2);
plot(f,M)
hold on
end

%plotchebymag.m
function [res]=plotchebymag(A,fc,epsilon)
figure
subplot(2,1,1)
%Magnitude response of the Chebyshev filter
for N=3:2:11
M=[];
for f=0:0.01:5;
M=[M A./(1+(epsilon^2)*CN(f/fc,N)^2)^(1/2)];
end
plot(0:0.01:5,M)
hold on
end
```

```
for N=2:2:11
M=[];
for f=0:0.01:5;
M=[M A./(1+(epsilon^2)*CN(f/fc,N)^2)^(1/2)];
end
plot(0:0.01:5,M,'r')
hold on
end
subplot(2,1,2)
for N=3:2:11
M=[];
for f=0:0.01:2;
M=[M A./(1+(epsilon^2)*CN(f/fc,N)^2)^(1/2)];
end
plot(0:0.01:2,M)
hold on
end

for N=2:2:11
M=[];
for f=0:0.01:2;
M=[M A./(1+(epsilon^2)*CN(f/fc,N)^2)^(1/2)];
end
plot(0:0.01:2,M,'r')
hold on
end
```

```
%CN.m
function [res]=CN(f,N)
switch f<1
    case 0
        res=cos(N*acos(f));
    case 1
        res=cosh(N*acosh(f));
end
```

```
%butterworthorder.m
function [N]=butterworthorder(A,wc,ws,m)
%Let the maximum frequency content is set as 10000 Hz
%A is the magnitude at w=0
%wc is the cut-off frequency at which
%the magnitude is A/sqrt(2)in rad/sec
%ws is the stop band cutoff frequency
%(in rad/sec) at which the magnitude expected is lesser than m
N=log(((A^2)/(m^2))-1)/(2*log(ws/wc));
N=ceil(N);
fc=wc/(2*pi);
f=0:1:10000;
M=A./(1+(f/fc).^(2*N)).^(1/2);
figure
plot(f,M)
%digitalbutterworth.m
function [NUM,DEN,H]=digitalbutterworth(A,wc,ws,m,Fs,option)
%option 1: Impulse-invariant technique
%option 2: Bilinear transformation technique
```

```
switch option
    case 1
[N]=butterworthorder(A,wc,ws,m);
N
Ts=1/Fs;
order=mod(N,2);
if(order==0)
    N1=N;
else
    N1=N-1;
end
b=0;
for k=1:1:(N1/2)
b(k)=2*sin((2*k-1)*pi/(2*N))
end
Ck=1;
Bk=(A)^(2/N);
B0=1;
c0=1;
%Converting s domain to z-domain
if(N~=1)
for k=1:1:length(b)
[NU,DE]=impulses2z(Bk,Ck,b(k),wc,Fs)
res1{k}=NU;
res2{k}=DE;
end
H=1;
NUM1=res1;
DEN1=res2;
for k=1:1:(N1/2)
[H1,W]=freqz(NUM1{k},DEN1{k});
H=H.*H1;
end
end
if(N==1)
    H=1;
end
if(order==1)
[H2,W]=freqz([B0*wc],[1 -exp(-c0*wc*Ts)])
H=H.*H2;
end
H=abs(H)/max(abs(H))*A;
figure
plot((W*Fs)/(2*pi),H)

if(N==1)
    NUM{1}=[B0*wc];
    DEN{1}=[1 -exp(-c0*wc*Ts)];
else
NUM=NUM2;
DEN=DEN2;
end
```

```
case 2
%Frequency prewarping
%Needs to design the digital filter with cutoff frequency
wcd=(wc/Fs);
wsd=(ws/Fs);
Ts=1/Fs;
pwc=(2/Ts)*tan(wcd/2);
pws=(2/Ts)*tan(wsd/2);
[N]=butterworthorder(A,pwc,pws,m);
N
order=mod(N,2);
if(order==0)
    N1=N;
else
    N1=N-1;
end
b=0;
for k=1:1:(N1/2)
b(k)=2*sin((2*k-1)*pi/(2*N));
end
Ck=1;
Bk=(A)^(2/N);
B0=1;
c0=1;
if(N~=1)
for k=1:1:length(b)
    [NU,DE]=bilinears2z(Bk,Ck,b(k),pwc,Fs);
    res1{k}=NU;
    res2{k}=DE;
end
NUM2=res1;
DEN2=res2;
H=1;
for k=1:1:(N1/2)
[H1,W]=freqz(NUM2{k},DEN2{k});
H=H.*H1;
end
else
H=1;
end

if(order==1)
[H2,W]=freqz([B0*pwc*Ts B0*pwc*Ts],[(2+c0*pwc*Ts) -2+c0*pwc*Ts]);
H=H.*H2;
end
H=abs(H)/max(abs(H))*A;
figure
plot((W*Fs)/(2*pi),H)

if(N==1)
    NUM{1}=[B0*pwc*Ts B0*pwc*Ts];
    DEN{1}=[(2+c0*pwc*Ts) -2+c0*pwc*Ts];
else
NUM=NUM2;
DEN=DEN2;
end

end
```

```
%impulses2z.m
function [NUM,DEN]=impulses2z(Bk,Ck,bk,wc,Fs)
Ts=1/Fs;
vector=[1 bk*wc Ck*(wc^2)];
[p]=roots(vector);
NUM=[0 (exp(p(1)*Ts)-exp(p(2)*Ts))];
NUM=NUM*Bk*(wc^2)/(p(1)-p(2));
DEN=conv([1 -1*exp(p(1)*Ts)],[1 -1*exp(p(2)*Ts)]);
```

```
%bilinears2z.m
function [NUM,DEN]=bilinears2z(Bk,Ck,bk,wc,Fs)
Ts=1/Fs;
NUM=[Bk*(wc^2)*(Ts^2) 2*Bk*(wc^2)*(Ts^2) Bk*(wc^2)*(Ts^2)];
DEN=[4-2*bk*wc*Ts+Ck*(wc^2)*(Ts^2) ...
-8+2*Ck*(wc^2)*(Ts^2) 4+2*bk*wc*Ts+Ck*(wc^2)*(Ts^2)];
```

```
%digitalchebyshev.m
function [NUM,DEN,H]=digitalchebyshev(A,R,wp,ws,m,Fs,option)
%option 1: Impulse invariant technique
%option 2: Bilinear transformation technique
switch option
    case 1
[N]=chebyshevorder(A,R,wp,m,ws)
N
Ts=1/Fs;
order=mod(N,2);
if(order==0)
    N1=N;
else
    N1=N-1;
end
epsilon=sqrt(R/(A-R));
t=(((1/epsilon^2)+1)^(1/2)+(1/epsilon));
Y= (1/2)*(t^(1/N)-t^(-1/N));
b=0;
for k=1:1:(N1/2)
b(k)=2*Y*sin((2*k-1)*pi/(2*N));
C(k)=Y^2+(cos((2*k-1)*pi/(2*N)))^(2);
end
Bk=(A)^(2/N);
B0=Bk;
C0=Y;
wc=wp;
%Converting s domain to z domain
if(N~=1)
for k=1:1:length(b)
[NU,DE]=impulses2z(Bk,C(k),b(k),wc,Fs)
res1{k}=NU;
res2{k}=DE;
end
H=1;
NUM1=res1;
DEN1=res2;
for k=1:1:(N1/2)
[H1,W]=freqz(NUM1{k},DEN1{k});
H=H.*H1;
```

```
end
else
H=1;
end


if(order==1)
[H2,W]=freqz([B0*wc],[1 -exp(-C0*wc*Ts)]);
H=H.*H2;
end
H=abs(H)/max(abs(H))*A;
figure
plot((W*Fs)/(2*pi),H)

if(N==1)
    NUM{1}=[B0*wc];
    DEN{1}=[1 -exp(-C0*wc*Ts)];
else
NUM=NUM1;
DEN=DEN1;
end



case 2
%Frequency prewarping
%Needs to design the digital filter with cutoff frequency
wpd=(wp/Fs);
wsd=(ws/Fs);
Ts=1/Fs;
pwp=(2/Ts)*tan(wpd/2);
pws=(2/Ts)*tan(wsd/2);
[N]=chebyshevorder(A,R,pwp,m,pws)
N
order=mod(N,2);
if(order==0)
    N1=N;
else
    N1=N-1;
end
epsilon=sqrt(R/(A-R));
t=(((1/epsilon^2)+1)^(1/2)+(1/epsilon));
Y= (1/2)*(t^(1/N)-t^(-1/N));
b=0;
for k=1:1:(N1/2)
b(k)=2*Y*sin((2*k-1)*pi/(2*N));
C(k)=Y^2+(cos((2*k-1)*pi/(2*N)))^(2);
end
Bk=(A)^(2/N);
B0=Bk;
C0=Y;
pwc=pwp;

if(N~=1)
for k=1:1:length(b)
    [NU,DE]=bilinears2z(Bk,C(k),b(k),pwc,Fs);
    res1{k}=NU;
    res2{k}=DE;
end
NUM2=res1;
```

```
DEN2=res2;
H=1;
for k=1:1:(N1/2)
[H1,W]=freqz(NUM2{k},DEN2{k});
H=H.*H1;
end
else
    H=1;
end


if(order==1)
[H2,W]=freqz([B0*pwc*Ts B0*pwc*Ts],[(2+C0*pwc*Ts) -2+C0*pwc*Ts]);
H=H.*H2;
end
H=abs(H)/max(abs(H))*A;
figure
plot((W*Fs)/(2*pi),H)


if(N==1)
    NUM{1}=[B0*pwc*Ts B0*pwc*Ts];
    DEN{1}=[(2+C0*pwc*Ts) -2+C0*pwc*Ts];
else
NUM=NUM2;
DEN=DEN2;
end

end
```

```
%chebyshevorder.m
function [N]=chebyshevorder(A,R,wp,m,ws)
%Let the maximum frequency content is set as 10000 Hz
%R is the ripple width
%A/sqrt(2) is the amplitude expected at wp=wc in rad/sec
%The amplitude expected at stopband cutoff frequency ws in rad/sec is lesser
%than m

epsilon=sqrt(R/(A-R));
r=ws/wp;
C=sqrt(((1/(m^2))-1)/(epsilon^2));
N=ceil(acosh(C)/acosh(r));
fc=wp/(2*pi);
M=[];
for f=0:1:10000;
M=[M A./(1+(epsilon^2)*CN(f/fc,N)^2)^(1/2)];
end
figure
plot(0:1:10000,M)
```

2.2.6 *Comments on Fig. 2.7 and Fig. 2.8*

1. Figure 2.7a shows the intended magnitude response of the Butterworth low-pass filter, which is obtained by plotting (2.9) for the typical values of N and wc. Figure 2.7b shows the magnitude response of the actually designed Butterworth filter. This is obtained by mapping $H_a(s)$ (refer (2.7) and (2.8)) to $H(z)$, followed by computing the magnitude response of the transfer function $H(z)$. It is seen that the intended magnitude response and the magnitude response of the designed filter are almost identical.
2. Figure 2.8a shows the intended magnitude response of the Chebyshev low-pass filter, which is obtained by plotting the (2.10) for the typical values of N, wc, and ε. Figure 2.8c shows the magnitude response of the actually designed Chebyshev filter. This is obtained by mapping $H_a(s)$ (refer (2.7) and (2.8)) to $H(z)$, followed by computing the magnitude response of the transfer function $H(z)$. It is seen that the intended magnitude response and the magnitude response of the designed filter are almost identical.
3. For the bilinear transformation, we need to get the prewarped specification to design the intended low-pass filter that has the magnitude response as shown in Fig. 2.7a (Butterworth filter) and Fig. 2.8a (Chebyshev filter). The magnitude

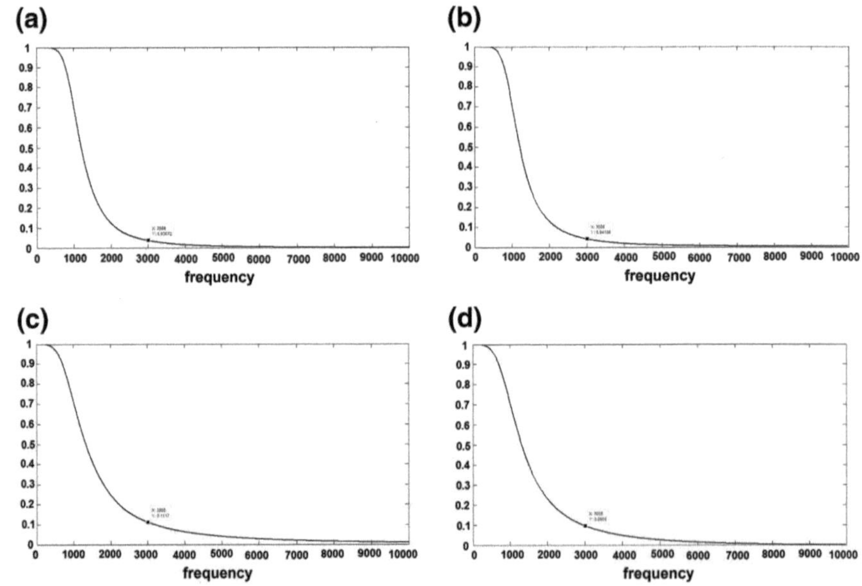

Fig. 2.7 Magnitude response of the designed Butterworth IIR low-pass filter (with magnitude response less than 0.1 at $fs = 3000$ Hz (stop band frequency) and 3dB cutoff at $fc = 500$ Hz (refer Sects. 2.2.2 and 2.2.3). The sampling frequency is $Fs = 20000$ Hz. **a** Intended low-pass filter. **b** Actually designed filter using impulse-invariant technique. **c** Specification after frequency prewarping. **d** Actual designed filter using bilinear transformation

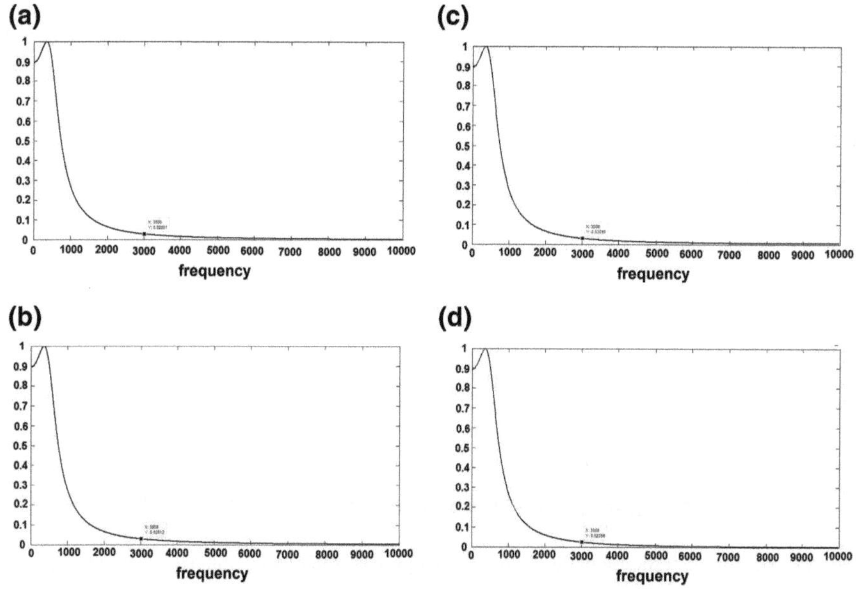

Fig. 2.8 Magnitude response of the designed Chebyshev IIR low-pass filter (with magnitude response less than 0.1 at $fs = 3000$ Hz (stop band frequency), $fc = 500$ Hz (refer Sects. 2.2.4 and 2.2.5) and $Ripplewidth(R) = 0.2$. The sampling frequency is $Fs = 20000$ Hz. **a** Intended low-pass filter. **b** Specification after frequency pre-warping. **c** Actually designed filter using impulse-invariant technique. **d** Actual designed filter using bilinear transformation

response of the IIR filter with the prewarped frequency specifications is shown in Fig. 2.7c (Butterworth filter) and Fig. 2.8b (Chebyshev filter) and the magnitude response of the actually designed IIR filter using bilinear transformation is shown in Fig. 2.7d (Butterworth filter) and Fig. 2.8d (Chebyshev filter). It is seen that amplitude of the magnitude response of the filter after transformation is lesser than the corresponding value in the prewarped specification. This helps in avoiding overlapping of spectrum.

2.2.7 *Design of High-Pass, Bandpass, and Band-Reject IIR Filter*

2.2.7.1 High-Pass Filter

Given the low-pass filter transfer function $H(e^{jw_d})$ with cutoff w_c radians, the high-pass filter is obtained as $H(e^{j(\pi - w_d)})$ with cutoff $\pi - w_c$. This is equivalent to replacing z with $-z$ in the z-transformation corresponding to LPF to obtain the HPF z-transform. Digital Butterworth high-pass filter using impulse invariant trans-

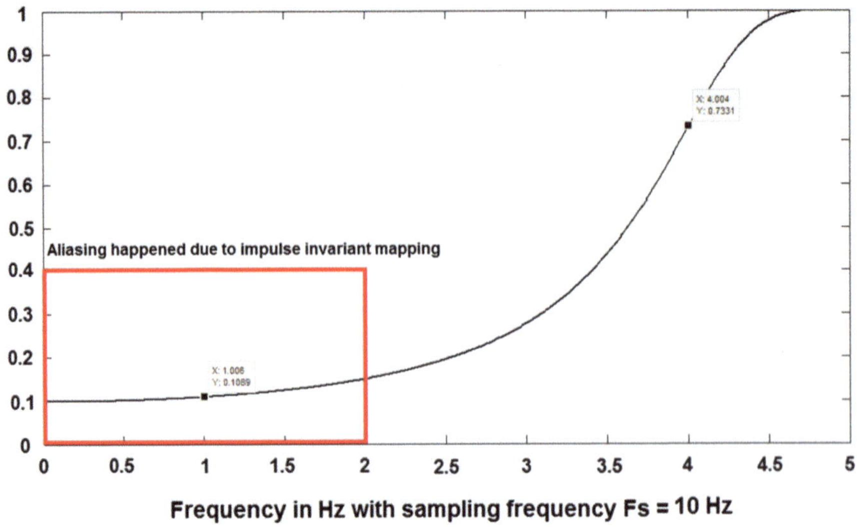

Fig. 2.9 Magnitude response of the Butterworth high-pass filter using impulse-invariant mapping

formation and bilinear transformation with pass band cutoff 8π rad/sec, stop band cutoff frequency 2π rad/sec, and sampling frequency $F_s = 10$ Hz is illustrated in Figs. 2.9 and 2.10, respectively. It is seen from Fig. 2.9 that the Aliasing occur at the lower frequencies. It is also noted that there exists nonzero amplitude at DC (0 Hz). This is the undesirable characteristics and hence impulse-invariant mapping is not usually used to design high-pass filter. This is circumvented using the bilinear transformation and is illustrated in Fig. 2.10.

```
%ButterworthHPFdemo.m
%Digital Butterworth high-pass filter using
%Impulse invariant and bilinear transformation with pass band cutoff
%2*pi*4 rad/sec, stop band cutoff frequency 2*pi*1 rad/sec
%magnitude at the stop band lesser than 0.1 and the sampling frequency 10 Hz
[NUM,DEN,H]=digitalbutterworthHPF(1,2*pi*4,2*pi*1,0.1,10,1)
[NUM,DEN,H]=digitalbutterworthHPF(1,2*pi*4,2*pi*1,0.1,10,1)
```

```
%digitalbutterworthHPF.m
function [NUM,DEN,H]=digitalbutterworthHPF(A,wc,ws,m,Fs,option)
wc=(pi-(wc/Fs))*Fs
ws=(pi-(ws/Fs))*Fs
%option 1: Impulse invariant technique
%option 2: Bilinear transformation technique
switch option
    case 1
[N]=butterworthorder(A,wc,ws,m);
N
Ts=1/Fs;
order=mod(N,2);
```

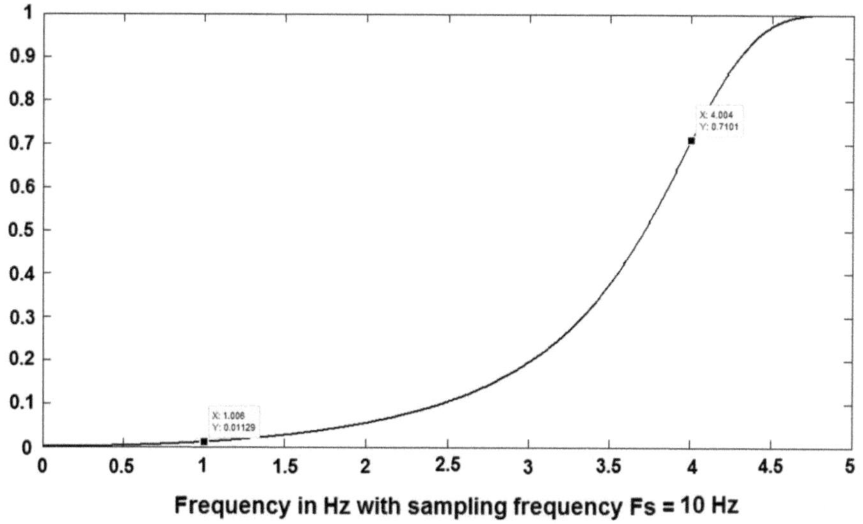

Fig. 2.10 Magnitude response of the Butterworth high-pass filter using bilinear transformation mapping

```
if(order==0)
    N1=N;
else
    N1=N-1;
end
b=0;
for k=1:1:(N1/2)
b(k)=2*sin((2*k-1)*pi/(2*N))
end
Ck=1;
Bk=(A)^(2/N);
B0=1;
c0=1;
%Converting s domain to z domain
if(N~=1)
for k=1:1:length(b)
[NU,DE]=impulses2z(Bk,Ck,b(k),wc,Fs)
L1=length(NU)
if(mod(L1,2)==1)
L=(L1+1)/2;
s1=[ones(1,L/2);zeros(1,L/2)]*2-1
s1=reshape(s1,1,size(s1,1)*size(s1,2))
s1=[s1 1];
else
L=L1/2;
s1=[ones(1,L);zeros(1,L)]*2-1
s1=reshape(s1,1,size(s1,1)*size(s1,2))
end
res1{k}=NU.*s1;

L2=length(DE);
if(mod(L2,2)==1)
```

```
L=(L2+1)/2;
s2=[ones(1,L/2);zeros(1,L/2)]*2-1;
s2=reshape(s2,1,size(s2,1)*size(s2,2));
s2=[s2 1];
else
L=L2/2;
s2=[ones(1,L);zeros(1,L)]*2-1;
s2=reshape(s2,1,size(s2,1)*size(s2,2));
end
res2{k}=DE.*s2
end
H=1;
NUM1=res1;
DEN1=res2;
for k=1:1:(N1/2)
[H1,W]=freqz(NUM1{k},DEN1{k})
H=H.*H1;
end
else
    H=1;
end

if(order==1)
[H2,W]=freqz([B0*wc],[1 exp(-c0*wc*Ts)]);
H=H.*H2;
end
H=abs(H)/max(abs(H))*A;
figure
plot((W*Fs)/(2*pi),H)

if(N==1)
NUM{1}=[B0*wc];
DEN{1}=[1 exp(-c0*wc*Ts)];
else
NUM=NUM1;
DEN=DEN1;
end


case 2
%Frequency prewarping
%Needs to design the digital filter with cutoff frequency
wcd=(wc/Fs);
wsd=(ws/Fs);
Ts=1/Fs;
pwc=(2/Ts)*tan(wcd/2);
pws=(2/Ts)*tan(wsd/2);
[N]=butterworthorder(A,pwc,pws,m);
N
order=mod(N,2);
if(order==0)
    N1=N;
else
    N1=N-1;
end
b=0;
for k=1:1:(N1/2)
b(k)=2*sin((2*k-1)*pi/(2*N));
end
Ck=1;
```

```
Bk=(A)^(2/N);
B0=1;
c0=1;
if(N~=1)
for k=1:1:length(b)
[NU,DE]=bilinears2z(Bk,Ck,b(k),pwc,Fs)
L1=length(NU);
if(mod(L1,2)==1)
L=(L1+1)/2;
s1=[ones(1,L/2);zeros(1,L/2)]*2-1;
s1=reshape(s1,1,size(s1,1)*size(s1,2));
s1=[s1 1];
else
L=L1/2;
s1=[ones(1,L/2);zeros(1,L/2)]*2-1;
s1=reshape(s1,1,size(s1,1)*size(s1,2));
end
res1{k}=NU.*s1;
L2=length(DE);
if(mod(L2,2)==1)
L=(L2+1)/2;
s2=[ones(1,L/2);zeros(1,L/2)]*2-1;
s2=reshape(s2,1,size(s2,1)*size(s2,2));
s2=[s2 1];
else
L=L2/2;
s2=[ones(1,L);zeros(1,L)]*2-1;
s2=reshape(s2,1,size(s2,1)*size(s2,2));
end
res2{k}=DE.*s2;
end
NUM2=res1;
DEN2=res2;
H=1;
for k=1:1:(N1/2)
[H1,W]=freqz(NUM2{k},DEN2{k});
H=H.*H1;
end
else
H=1;
end

if(order==1)
[H2,W]=freqz([B0*pwc*Ts -B0*pwc*Ts],[(2+c0*pwc*Ts) 2-c0*pwc*Ts]);
H=H.*H2;
end
H=abs(H)/max(abs(H))*A;
figure
plot((W*Fs)/(2*pi),H)

if(N==1)
NUM{1}=[B0*pwc*Ts -B0*pwc*Ts]
DEN{1}=[(2+c0*pwc*Ts) 2-c0*pwc*Ts]
else
NUM=NUM2;
DEN=DEN2;
end
end
```

```
%chebyshevHPFdemo.m
%Digital Chebyshev high-pass filter using
%Impulse invariant and Bilinear transformation with pass band cutoff
%2*pi*4 rad/sec, stop band cutoff frequency 2*pi*1 rad/sec, Ripple width 0.5
%magnitude at the stop band lesser than 0.1 and the sampling frequency 10 Hz
[NUM,DEN,H]=digitalchebyshevHPF(1,0.5,2*pi*4,2*pi*1,0.1,10,1)
[NUM,DEN,H]=digitalchebyshevHPF(1,0.5,2*pi*4,2*pi*1,0.1,10,2)
```

```
%digitalchebyshevHPF.m
function [NUM,DEN,H]=digitalchebyshevHPF(A,R,wp,ws,m,Fs,option)
wp=(pi-(wp/Fs))*Fs
ws=(pi-(ws/Fs))*Fs
%option 1: Impulse invariant technique
%option 2: Bilinear transformation technique
switch option
    case 1
[N]=chebyshevorder(A,R,wp,m,ws)
N
Ts=1/Fs;
order=mod(N,2);
if(order==0)
    N1=N;
else
    N1=N-1;
end
epsilon=sqrt(R/(A-R));
t=(((1/epsilon^2)+1)^(1/2)+(1/epsilon));
Y= (1/2)*(t^(1/N)-t^(-1/N));
b=0;
for k=1:1:(N1/2)
b(k)=2*Y*sin((2*k-1)*pi/(2*N));
C(k)=Y^2+(cos((2*k-1)*pi/(2*N)))^(2);
end
Bk=(A)^(2/N);
B0=Bk;
C0=Y;
wc=wp;
%Converting s domain to z domain
if(N~=1)
for k=1:1:length(b)
[NU,DE]=impulses2z(Bk,C(k),b(k),wc,Fs)
L1=length(NU)
if(mod(L1,2)==1)
L=(L1+1)/2;
s1=[ones(1,L/2);zeros(1,L/2)]*2-1
s1=reshape(s1,1,size(s1,1)*size(s1,2))
s1=[s1 1];
else
L=L1/2;
s1=[ones(1,L);zeros(1,L)]*2-1
s1=reshape(s1,1,size(s1,1)*size(s1,2))
end
res1{k}=NU.*s1;

L2=length(DE);
if(mod(L2,2)==1)
L=(L2+1)/2;
s2=[ones(1,L/2);zeros(1,L/2)]*2-1;
```

```
s2=reshape(s2,1,size(s2,1)*size(s2,2));
s2=[s2 1];
else
L=L2/2;
s2=[ones(1,L);zeros(1,L)]*2-1;
s2=reshape(s2,1,size(s2,1)*size(s2,2));
end
res2{k}=DE.*s2
end
H=1;
NUM1=res1;
DEN1=res2;
for k=1:1:(N1/2)
[H1,W]=freqz(NUM1{k},DEN1{k});
H=H.*H1;
end
else
H=1;
end
if(order==1)
[H2,W]=freqz([B0*wc],[1 exp(-C0*wc*Ts)]);
H=H.*H2;
end
H=abs(H)/max(abs(H))*A;
figure
plot((W*Fs)/(2*pi),H)

if(N==1)
NUM{1}=[B0*wc];
DEN{1}=[1 exp(-C0*wc*Ts)];
else
NUM=NUM1;
DEN=DEN1;
end



case 2
%Frequency prewarping
%Needs to design the digital filter with cutoff frequency
wpd=(wp/Fs);
wsd=(ws/Fs);
Ts=1/Fs;
pwp=(2/Ts)*tan(wpd/2);
pws=(2/Ts)*tan(wsd/2);
[N]=chebyshevorder(A,R,pwp,m,pws)
N
order=mod(N,2);
if(order==0)
    N1=N;
else
    N1=N-1;
end
epsilon=sqrt(R/(A-R));
t=(((1/epsilon^2)+1)^(1/2)+(1/epsilon));
Y= (1/2)*(t^(1/N)-t^(-1/N));
b=0;
for k=1:1:(N1/2)
b(k)=2*Y*sin((2*k-1)*pi/(2*N));
C(k)=Y^2+(cos((2*k-1)*pi/(2*N)))^(2);
```

```
end
Bk=(A)^(2/N);
B0=Bk;
C0=Y;
pwc=pwp;
if(N~=1)
for k=1:1:length(b)
    [NU,DE]=bilinears2z(Bk,C(k),b(k),pwc,Fs);
    L1=length(NU)
if(mod(L1,2)==1)
L=(L1+1)/2;
s1=[ones(1,L/2);zeros(1,L/2)]*2-1
s1=reshape(s1,1,size(s1,1)*size(s1,2))
s1=[s1 1];
else
L=L1/2;
s1=[ones(1,L);zeros(1,L)]*2-1
s1=reshape(s1,1,size(s1,1)*size(s1,2))
end
res1{k}=NU.*s1;

L2=length(DE);
if(mod(L2,2)==1)
L=(L2+1)/2;
s2=[ones(1,L/2);zeros(1,L/2)]*2-1;
s2=reshape(s2,1,size(s2,1)*size(s2,2));
s2=[s2 1];
else
L=L2/2;
s2=[ones(1,L);zeros(1,L)]*2-1;
s2=reshape(s2,1,size(s2,1)*size(s2,2));
end
res2{k}=DE.*s2;
end
NUM2=res1;
DEN2=res2;
H=1;
for k=1:1:(N1/2)
[H1,W]=freqz(NUM2{k},DEN2{k});
H=H.*H1;
end
else
H=1;
end


if(order==1)
[H2,W]=freqz([B0*pwc*Ts -B0*pwc*Ts],[(2+C0*pwc*Ts) 2-C0*pwc*Ts]);
H=H.*H2;
end
H=abs(H)/max(abs(H))*A;
figure
plot((W*Fs)/(2*pi),H)

if(N==1)
NUM{1}=[B0*pwc*Ts -B0*pwc*Ts ];
```

```
DEN{1}=[(2+C0*pwc*Ts) 2-C0*pwc*Ts ];
else
NUM=NUM2;
DEN=DEN2;
end

end
```

2.2.7.2 Bandpass Filter

Bandpass filter is obtained as the cascade of low-pass filter with cutoff frequency wc_2 and high-pass filter with cutoff frequency wc_1 (Figs. 2.11 and 2.12). The bandpass filter with $wc_1 = 2\pi$ rad/sec and $wc_2 = 8\pi$ rad/sec is illustrated in Fig. 2.13a–c (Butterworth filter) and Fig. 2.13d–f (Chebyshev filter) using bilinear transformation technique. It is constructed using the cascade connection of low-pass filter (with cutoff frequency $wc_2 = 8\pi$ rad/sec and stop band cutoff frequency $ws_2 = 2\pi 0.1\frac{F_s}{2}$ rad/sec), followed by the high-pass filter (with cutoff frequency $wc_1 = 2\pi$ rad/sec and $ws_1 = 2\pi 0.9\frac{F_s}{2}$ rad/sec). Impulse-invariant (lead to overlapping) is not usually chosen to design other than low-pass filter. Hence, illustration of bandpass filter using bilinear transformation is demonstrated.

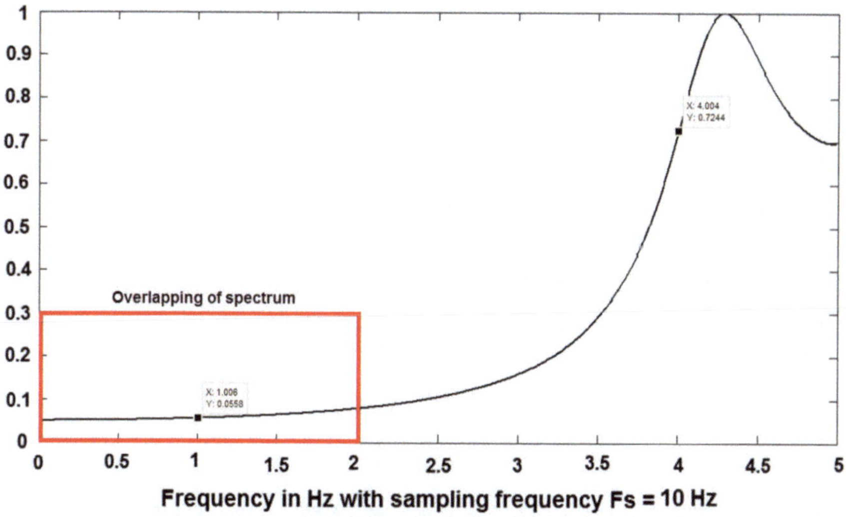

Fig. 2.11 Magnitude response of the Chebyshev high-pass filter using impulse-invariant mapping. It is seen that the magnitude is nonzero at $f = 0\,\text{Hz}$. This is due to overlapping of spectrum

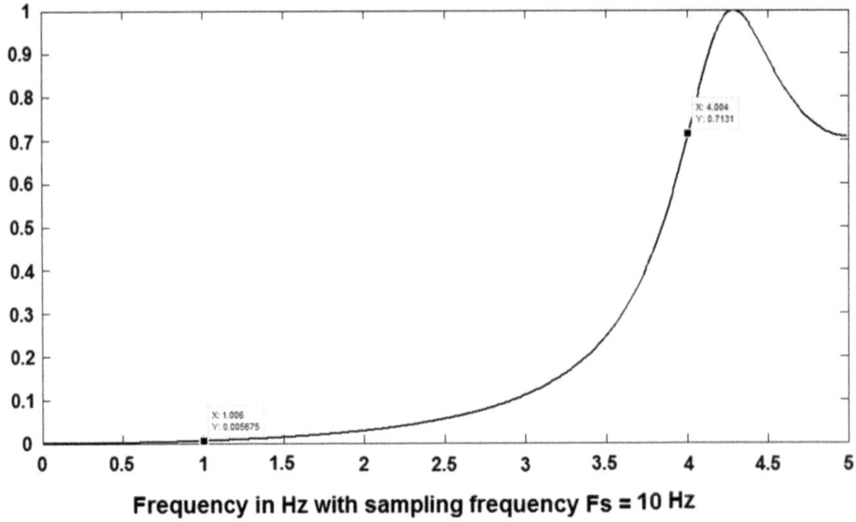

Fig. 2.12 Magnitude response of the Chebyshev high-pass filter using bilinear transformation mapping

```
%IIRBPFDEMO.m
A=1;
wc1=2*pi*1;
wc2=2*pi*4;
m=0.001;
Fs=10;
Ripple=0.5;

%Using Butterworth filter and impulse-invariant transformation
[NUM,DEN,H]=digitalBPF(1,Ripple,wc1,wc2,m,Fs,1,2);
figure;
plot(linspace(0,Fs/2,length(H)),abs(H));

%Using Chebyshevfilter and bilinear transformation
[NUM,DEN,H]=digitalBPF(1,Ripple,wc1,wc2,m,Fs,2,2);
figure
plot(linspace(0,Fs/2,length(H)),abs(H));
```

```
function [NUM,DEN,H]=digitalBPF(A,R,wc1,wc2,m,Fs,option1,option2)
%R is the ripple width used in case of Chebyshev filter
%A is the maximum amplitude of the filter
%H is the normalized magnitude response of the designed filter
%wc1 and wc2 are the cutoff frequencies in rad/sec
%option1:1->Butterworth 2->Chebyshev filter
%option2: 1->Impulse invariant 2->Bilinear
Fmax=Fs/2;
```

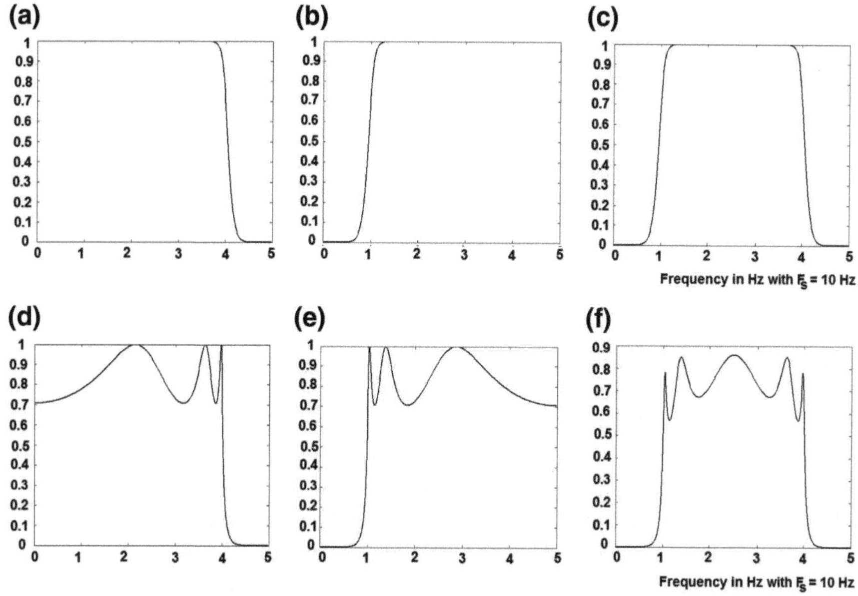

Fig. 2.13 Bandpass filter using bilinear transformation. **a** Butterworth low-pass filter. **b** Butterworth high-pass filter. **c** Corresponding Butterworth bandpass filter as the cascade of low-pass and high-pass filter. **d** Chebyshev low-pass filter. **e** Chebyshev high-pass filter. **f** Corresponding Chebyshev bandpass filter as the cascade of low-pass and high-pass filter

```
ws1=2*pi*0.1*(Fmax);
ws2=2*pi*0.9*(Fmax);
switch option1
    case 1
        switch option2
            case 1
[N1,D1,H1]=digitalbutterworth(A,wc2,ws2,m,Fs,1);
[N2,D2,H2]=digitalbutterworthHPF(A,wc1,ws1,m,Fs,1);
            case 2
[N1,D1,H1]=digitalbutterworth(A,wc2,ws2,m,Fs,2);
[N2,D2,H2]=digitalbutterworthHPF(A,wc1,ws1,m,Fs,2);
        end
    case 2
        switch option2
            case 1
[N1,D1,H1]=digitalchebyshev(A,R,wc2,ws2,m,Fs,1);
[N2,D2,H2]=digitalchebyshevHPF(A,R,wc1,ws1,m,Fs,1);
            case 2
[N1,D1,H1]=digitalchebyshev(A,R,wc2,ws2,m,Fs,2);
[N2,D2,H2]=digitalchebyshevHPF(A,R,wc1,ws1,m,Fs,2);
        end
end

temp1=1;
temp2=1;
temp3=1;
```

```
for i=1:1:length(N1)
temp1=conv(temp1,N1{i})
temp2=conv(temp2,D1{i})
end
for i=1:1:length(N2)
temp1=conv(temp1,N2{i})
temp2=conv(temp2,D2{i})
end
NUM=temp1;
DEN=temp2;
H=H1.*H2;
```

2.2.7.3 Band-reject Filter

Band-reject filter is obtained as the parallel connection of low-pass filter with cutoff frequency wc_1 and the high-pass filter with cutoff frequency wc_2. The band-reject filter with $wc_1 = 2\pi$ rad/sec and $wc_2 = 8\pi$ rad/sec is illustrated in Fig. 2.14a–c (Butterworth filter) and Fig. 2.14d–f (Chebyshev filter) using bilinear transformation technique. It is constructed using the parallel connection of low-pass filter (with cutoff frequency $wc_1 = 2\pi$ rad/sec and stop band cutoff frequency $ws_1 = 2\pi 0.9\frac{F_s}{2}$), followed by the high-pass filter (with cutoff frequency $wc_2 = 2\pi$ rad/sec and $ws_2 = 2\pi 0.1\frac{F_s}{2}$). Impulse-invariant (lead to overlapping) is not usually chosen to design other than low-pass filter. Hence, realization using the bilinear transformation is demonstrated.

```
%IIRBRFdemo.m
A=1;
wc1=2*pi*1;
wc2=2*pi*4;
m=0.001;
Fs=10;
Ripple=0.5;

%Using Butterworth filter and impulse-invariant transformation
[NUM,DEN,H]=digitalBRF(1,Ripple,wc1,wc2,m,Fs,1,2);
figure;
plot(linspace(0,Fs/2,length(H{1})),(1/2)*(abs(H{1})+abs(H{2})));

%Using Chebyshev filter and bilinear transformation
[NUM,DEN,H]=digitalBRF(1,Ripple,wc1,wc2,m,Fs,2,2);
figure
plot(linspace(0,Fs/2,length(H{1})),(1/2)*(abs(H{1})+abs(H{2})));


%digitalBRF.m
function [NUM,DEN,H]=digitalBRF(A,R,wc1,wc2,m,Fs,option1,option2)
%option1:1->Butterworth 2->Chebyshev filter
%option2: 1->Impulse invariant 2->Bilinear
%R is the ripple width used in case of Chebyshev filter
Fmax=Fs/2;
ws1=2*pi*0.9*(Fmax);
ws2=2*pi*0.1*(Fmax);
switch option1
    case 1
```

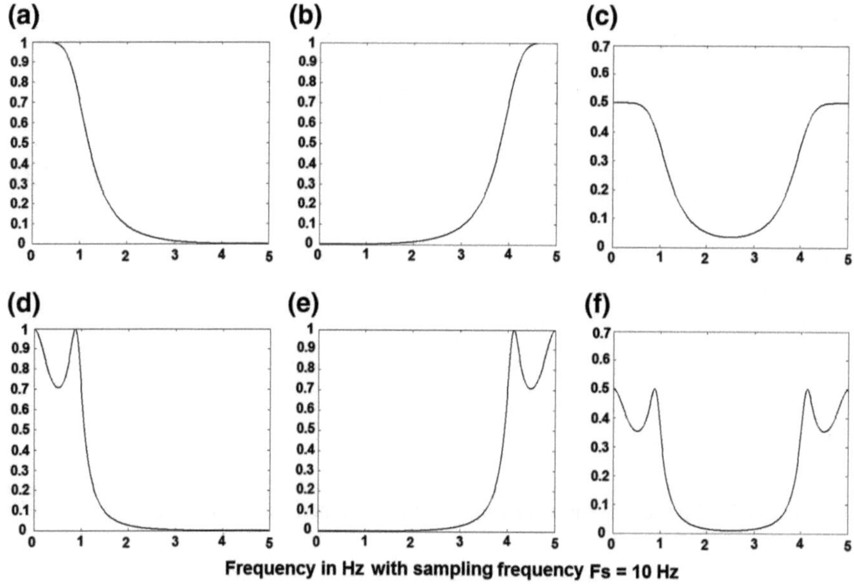

Fig. 2.14 Band-reject filter using bilinear transformation. **a** Butterworth low-pass filter. **b** Butterworth high-pass filter. **c** Corresponding band-reject filter as the parallel summation of low-pass and high-pass filter. **d** Chebyshev low-pass filter. **e** Chebyshev high-pass filter. **f** Corresponding Chebyshev band-reject filter as the parallel summation of low-pass and high-pass filter

```
          switch option2
              case 1
[N1,D1,H1]=digitalbutterworth(A,wc1,ws1,m,Fs,1);
figure
[N2,D2,H2]=digitalbutterworthHPF(A,wc2,ws2,m,Fs,1);
              case 2
[N1,D1,H1]=digitalbutterworth(A,wc1,ws1,m,Fs,2);
[N2,D2,H2]=digitalbutterworthHPF(A,wc2,ws2,m,Fs,2);
          end
      case 2
          switch option2
              case 1
[N1,D1,H1]=digitalchebyshev(A,R,wc1,ws1,m,Fs,1);
[N2,D2,H2]=digitalchebyshevHPF(A,R,wc2,ws2,m,Fs,1);
              case 2
[N1,D1,H1]=digitalchebyshev(A,R,wc1,ws1,m,Fs,2);
[N2,D2,H2]=digitalchebyshevHPF(A,R,wc2,ws2,m,Fs,2);
          end
end

NUM=N1;
DEN=D1;
H{1}=H1;
H{2}=H2;
```

2.3 Realization

Let the transfer function of the typical IIR filter is given as follows:

$$H(z) = \frac{a_0 + a_1 z^{-1} + a_2 z^{-2} + \cdots a_p z^{-p}}{b_0 + b_1 z^{-1} + b_2 z^{-3} + \cdots b_q z^{-q}}. \tag{2.11}$$

Realization of the IIR filter is the method of obtaining the output sequence $y(n)$ corresponding to the input sequence $x(n)$ to the linear IIR filter $h(n)$. This is done as follows.

2.3.1 *Direct Form 1*

Let $X(z)$, $Y(z)$ be the z-transformation of the sequence $x(n)$ and $y(n)$, respectively:

$$\frac{Y(z)}{X(z)} = \frac{a_0 + a_1 z^{-1} + a_2 z^{-2} + a_3 z^{-3} + \cdots a_p z^{-p}}{b_0 + b_1 z^{-1} + b_2 z^{-2} + b_3 z^{-3} + \cdots + b_q z^{-q}}. \tag{2.12}$$

Taking inverse z-transformation, we get the following difference equations:

$$\begin{aligned} y(n) = {} & \frac{a_0}{b_0} x(n) + \frac{a_1}{b_0} x(n-1) + \frac{a_2}{b_0} x(n-2) + \cdots + \frac{a_p}{b_0} x(n-p) \\ & - \frac{b_1}{b_0} y(n-1) - \frac{b_2}{b_0} y(n-2) - \cdots - \frac{b_q}{b_0} y(n-q). \end{aligned}$$

2.3.2 *Direct Form 2*

Let $\frac{Y(z)}{X(z)} = \frac{Y(z)}{W(z)} \frac{W(z)}{X(z)}$, $\frac{Y(z)}{W(z)} = a_0 + a_1 z^{-1} + a_2 z^{-2} + \cdots a_p z^{-p}$ and $\frac{W(z)}{X(z)} = \frac{1}{b_0 + b_1 z^{-1} + b_2 z^{-2} + \cdots b_q z^{-q}}$. Taking inverse z-transformation, we get the following difference equations:

$$\begin{aligned} w(n) &= -\frac{1}{b_0} x(n) - \frac{b_1}{b_0} w(n-1) - \frac{b_2}{b_0} w(n-2) \\ y(n) &= a_0 w(n) + a_1 w(n-1) + a_2 w(n-2). \end{aligned}$$

We see that to realize IIR filter using Direct form I, we need number of $(p+q)$ number of taps. But to realize using Direct form II, we need only $max(p, q)$ number of taps at the cost of time required for the computation.

2.3.3 Illustration

Consider the input signal $x(t) = \sum_{k=1}^{k=3} A_k sin(2\pi f_k t)$ is sampled using the sampling frequency Fs to obtain the discrete sequence $x(n) = \sum_{k=1}^{k=3} sin(2\pi f_k n Ts)$. The digital impulse-invariant Butterworth IIR filter is designed to filter f_3 as given below.

1. Let $A_1 = 1$, $A_2 = 1$ and $A_3 = 1$, $f_1 = 10$, $f_2 = 15$ and $f_3 = 200$.
2. The specification is obtained as follows: Butterworth low-pass filter is designed with $A = 1$, $wc = 2 * pi * 30$, $ws = 2 * pi * 100$, $Fs = 500$ and the amplitude is lesser than 0.1 at ws.
3. The transfer function of the filter is obtained as

$$H(z) = \frac{53.7905Z^{-1}}{1 - 1.4779Z^{-1} + 0.5868Z^{-2}}. \tag{2.13}$$

4. Direct form 1: For the input sequence $x(n)$, the corresponding output sequence is obtained as follows: $y(n) = 53.79x(n-1) - 1.477y(n-1) - 0.5868y(n-2)$.
5. Direct form 2: For the input sequence $x(n)$, the corresponding output sequence is obtained as follows: $w(n) = -x(n) + 1.477w(n-1) - 0.5868w(n-2)$, $y(n) = 53.79w(n-1)$.
6. The number of taps needed for realization of the filter is 3 for Direct form I (DF1) and 2 for Direct form II (DF2). The elapsed time required for DF1 and DF2 realization is given as 0.005619 and 0.009127 s, respectively.
7. Figure 2.15 illustrates the realization of IIR filter using Direct form I and are identical with that of the magnitude response realized using Direct form II. Figure 2.16 illustrates the magnitude response of IIR filter corresponding to the transfer function (2.13).

```
%realizeiir.m
A1=1;
A2=1;
A3=1;
f1=10;
f2=15;
f3=200;
A=1;
wc=2*pi*30;
ws=2*pi*100;
Fs=500;
Ts=1/Fs;
m=0.1;
n=0:1:1000;
S=A1*sin(2*pi*f1*n*Ts)+A2*sin(2*pi*f2*n*Ts)+A3*sin(2*pi*f3*n*Ts);
[NUM,DEN,H]=digitalbutterworth(A,wc,ws,m,Fs,1);
NUM{1}
DEN{1}
temp1=1;
temp2=1;
for k=1:1:length(NUM)
temp1=conv(temp1,NUM{k});
```

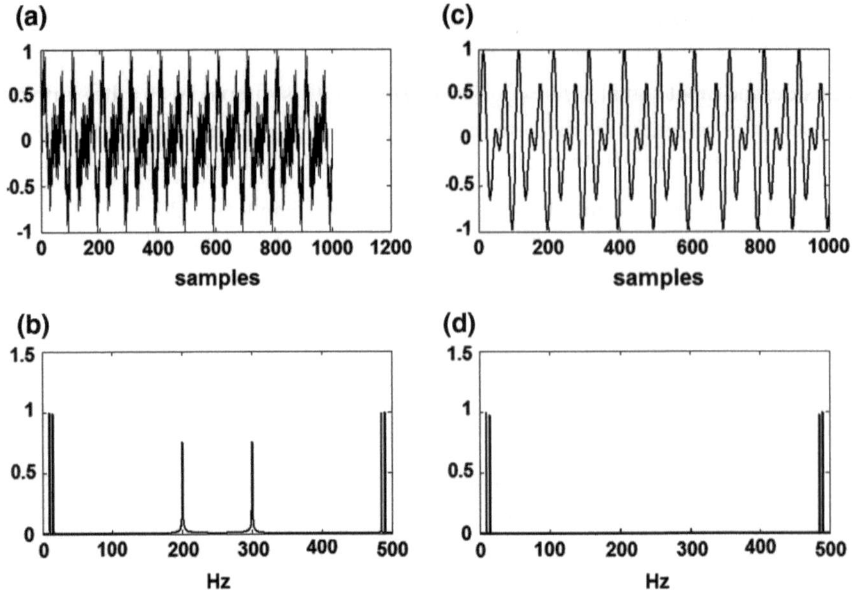

Fig. 2.15 Demonstration on the Direct form I realization of IIR filter using **a** input signal, **b** Filtered signal, **c** spectrum of the input signal, **d** spectrum of the filtered signal

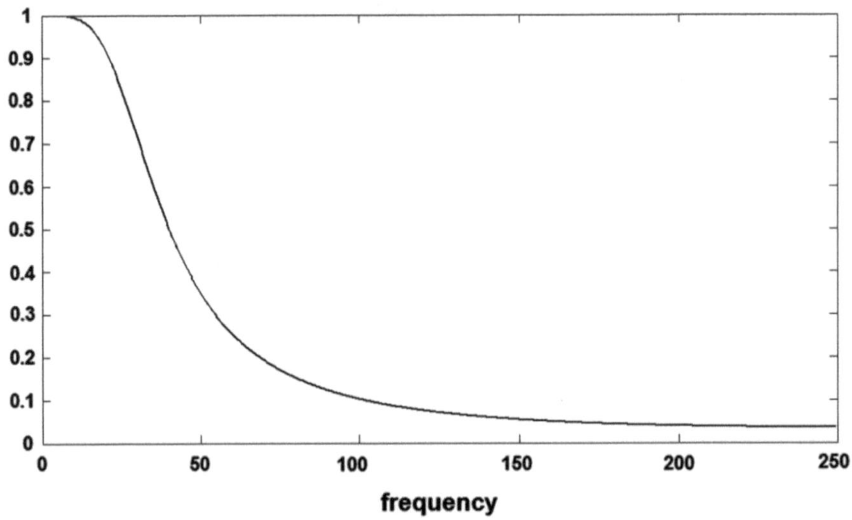

Fig. 2.16 Magnitude response of the IIR filter used to filter the input signal (refer Fig. 2.15)

```
end
for k=1:1:length(DEN)
temp2=conv(temp2,DEN{k});
end
temp1=temp1+eps;
[H,W]=freqz(temp1,temp2);
figure
plot(Fs*W/(2*pi),abs(H)/max(abs(H)))
%Direct form I realization
y=zeros(1,length(temp2)+1);
tic
for n=length(y)+1:1:1000
temp=0;
for r=0:1:length(temp1)-1
temp=temp+temp1(r+1)*S(n-r);
end
for s=1:1:length(temp2)-1
temp=temp-1*temp2(s+1)*y(n-s);
end
temp=temp/temp2(1);
y=[y temp];
end
toc
S=S/max(S);
y=y/max(y);
FRS=abs(fft(S))/max(abs(fft(S)));
FRy=abs(fft(y))/max(abs(fft(y)));
figure
subplot(2,2,1)
plot(S)
subplot(2,2,2)
plot(y)
subplot(2,2,3)
plot(linspace(0,Fs,length(S)),FRS)
subplot(2,2,4)
plot(linspace(0,Fs,length(y)),FRy)

%Directform II realization
M=max(length(temp1),length(temp2));
y=zeros(1,M+1);
w=zeros(1,M+1);
tic
for n=length(w):1:1000
temp=0;
temp=S(n);
for r=1:1:length(temp2)-1
temp=temp-temp2(r+1)*w(n-r);
end
w(n)=temp/temp2(1);
temp=0;
for s=0:1:length(temp1)-1
temp=temp+temp1(s+1)*w(n-s);
end

y=[y temp];
end
toc
S=S/max(S);
y=y/max(y);
```

```
FRS=abs(fft(S))/max(abs(fft(S)));
FRy=abs(fft(y))/max(abs(fft(y)));
figure
subplot(2,2,1)
plot(S)
subplot(2,2,2)
plot(y)
subplot(2,2,3)
plot(linspace(0,Fs,length(S)),FRS)
subplot(2,2,4)
plot(linspace(0,Fs,length(y)),FRy)
```

Chapter 3
Finite Impulse Response Filter (FIR Filter)

The impulse response of the linear phase FIR filter satisfies the condition $h(n) = \pm h(N-n-1)$, with $h(n)$ nonzero for $n = 0 \cdots (N-1)$. The magnitude and the phase response of linear phase FIR filter are summarized below.

1. Type 1: N is odd (consider $N = 5$) with $h(n) = h(N-1-n)$

$$H(e^{jw_d}) = h(0) + h(1)e^{-jw_d} + h(2)e^{-j2w_d} + h(3)e^{-j3w_d} + h(4)e^{-j4w_d}$$
$$\Rightarrow H(e^{jw_d}) = e^{-j2w_d}(h(0)e^{j2w_d} + h(1)e^{jw_d} + h(2) + h(3)e^{-jw_d} + h(4)e^{-j2w_d})$$
$$\Rightarrow e^{-j2w_d}(h(0)e^{j2w_d} + h(1)e^{jw_d} + h(2) + h(1)e^{-jw_d} + h(0)e^{-j2w_d})$$
$$\Rightarrow e^{-j2w_d}(h(2) + 2h(0)cos(2w_d) + 2h(1)cos(w_d))$$

2. Type 2: N is odd (consider $N = 5$)with $h(n) = -h(N-1-n)$ with $h(\frac{N-1}{2}) = 0$

$$H(e^{jw_d}) = h(0) + h(1)e^{-jw_d} + h(2)e^{-j2w_d} + h(3)e^{-j3w_d} + h(4)e^{-j4w_d}$$
$$\Rightarrow H(e^{jw_d}) = e^{-j2w_d}(h(0)e^{j2w_d} + h(1)e^{jw_d} + h(2) + h(3)e^{-jw_d} + h(4)e^{-j2w_d})$$
$$\Rightarrow e^{-j2w_d}(h(0)e^{j2w_d} + h(1)e^{jw_d} + h(2) - h(1)e^{-jw_d} - h(0)e^{-j2w_d})$$
$$\Rightarrow e^{-j2w_d}e^{j\frac{\pi}{2}}(2h(0)sin(2w_d) + 2h(1)sin(w_d))$$

3. Type 3: N is even (consider $N = 4$) with $h(n) = h(N-1-n)$
   Consider the transfer function of the Type 3 FIR filter with $N = 4$.

$$H(e^{jw_d}) = h(0) + h(1)e^{-jw_d} + h(2)e^{-j2w_d} + h(3)e^{-j3w_d}$$
$$\Rightarrow H(e^{jw_d}) = e^{-j\frac{3w_d}{2}}(h(0)e^{j\frac{3w_d}{2}} + h(1)e^{j\frac{w_d}{2}} + h(1)e^{-j\frac{w_d}{2}} + h(0)e^{-j\frac{3w_d}{2}})$$
$$\Rightarrow e^{-j\frac{3w_d}{2}}(2h(0)cos(\frac{w_d}{2}) + 2h(1)cos(\frac{3w_d}{2}))$$

4. Type 4: N is even (consider $N = 4$) with $h(n) = -h(N-1-n)$

E.S. Gopi, *Multi-Disciplinary Digital Signal Processing*,
DOI 10.1007/978-3-319-57430-1_3

$$H(e^{jw_d}) = h(0) + h(1)e^{-jw_d} + h(2)e^{-j2w_d} + h(3)e^{-j3w_d}$$

$$\Rightarrow H(e^{jw_d}) = e^{-j\frac{3w_d}{2}}(h(0)e^{j\frac{3w_d}{2}}h(1)e^{j\frac{w_d}{2}} - h(1)e^{-j\frac{w_d}{2}} - h(0)e^{-j\frac{3w_d}{2}})$$

$$\Rightarrow e^{-j\frac{3w_d}{2}}(2jh(0)sin(\frac{w_d}{2}) + 2jh(1)sin(\frac{3w_d}{2}))$$

$$\Rightarrow e^{-j\frac{3w_d}{2}}e^{j\frac{\pi}{2}}(2h(0)sin(\frac{w_d}{2}) + 2h(1)sin(\frac{3w_d}{2}))$$

3.1 Demonstration of Four Types of FIR Filter

Given the sequence of impulse response of the FIR filter be represented as $h(n)$, the magnitude and the phase response of the filter is obtained as

$$M = abs(\sum_{n=0}^{N-1} h(n)e^{-jw_d n})$$

$$P = angle(\sum_{n=0}^{N-1} h(n)e^{-jw_d n})$$

In particular, the generalized expressions for the four types of FIR filter with an arbitrary value of N are summarized below. Let $\tau = \frac{N-1}{2}$. The transfer function of the ith type of FIR filter is obtained as $M_i e^{-jP_i}$, where

$$M_1 = h(0) + \sum_{n=1}^{\frac{N-1}{2}} 2h(n)cos((\tau - n + 1)w_d) \tag{3.1}$$

$$P_1 = -\tau w_d \tag{3.2}$$

$$M_2 = \sum_{n=1}^{\frac{N-1}{2}} 2h(n)sin((\tau - n + 1)w_d) \tag{3.3}$$

$$P_2 = -\tau w_d + \frac{\pi}{2} \tag{3.4}$$

$$M_3 = h(0) + \sum_{n=1}^{\frac{N-1}{2}} 2h(n)cos((\tau - n + 1)w_d) \tag{3.5}$$

$$P_3 = -\tau w_d \tag{3.6}$$

$$M_4 = \sum_{n=1}^{\frac{N-1}{2}} 2h(n)sin((\tau - n + 1)w_d) \tag{3.7}$$

$$P_4 = -\tau w_d + \frac{\pi}{2} \tag{3.8}$$

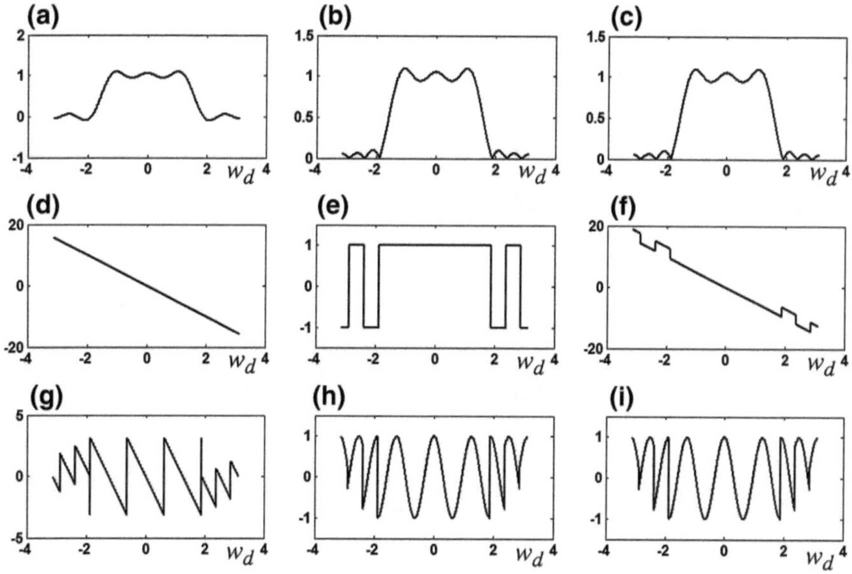

Fig. 3.1 Demonstration of Type 1 FIR filter. **a** M_1, **b** $abs(M_1)$, **c** $abs(M)$ computed for Type 1 filter, **d** P_1, **e** S = sign of M_1, **f** $PH_1 = (S-1)\frac{-\pi}{2} + P_1$, **g** P computed for Type 1 filter, **h** $cos(P)$, **i** $cos(PH_1)$

Figures 3.1, 3.2, 3.3 and 3.4 demonstrates the computation of magnitude and the phase response of the four types of FIR filter with the following specifications.

- Type 1: $N = 11, h(n) = \frac{sin(w_c(n-\frac{(N-1)}{2}))}{\pi(n-\frac{(N-1)}{2})}$ for $n = 0 \cdots \frac{(N-1)}{2}, h(n) = h(N-n-1)$
- Type 2: $N = 11$, $h(n) = \frac{sin(w_c(n-\frac{(N-1)}{2}))}{\pi(n-\frac{(N-1)}{2})}$ for $n = 0 \cdots \frac{(N-1)}{2}$, $h(n) = -h(N-n-1)$, $h(\frac{N-1}{2}) = 0$
- Type 3: $N = 10$, $h(n) = \frac{sin(w_c(n-\frac{(N-1)}{2}))}{\pi(n-\frac{(N-1)}{2})}$ for $n = 0 \cdots \frac{N}{2}$, $h(n) = h(N-n-1)$
- Type 4: $N = 10$, $h(n) = \frac{sin(w_c(n-\frac{(N-1)}{2}))}{\pi(n-\frac{(N-1)}{2})}$ for $n = 0 \cdots \frac{N}{2}$, $h(n) = -h(N-n-1)$.

The details on the Pole-zero plot (refer Fig. 3.5) are summarized below.

1. The generalized expression for the polynomial $H(z)$ is given as $H(z) = h(0) + h(1)z^{-1} + h(2)z^{-2} + \cdots + h(N-1)z^{-(N-1)}$.
2. For the real coefficients $[h(0)\, h(1) \cdots h(N-1)]$, the roots of the polynomial $H(z)$ are available as the complex conjugate pair.
3. For Type 2 with $N = 5$, we get the polynomial as $H(z) = h(0) + h(1)z^{-1} + h(2)z^{-2} - h(1)z^{-3} - h(0)z^{-4}$ with $h(2) = 0$. From the expression, we observe that $H(z) = 0$ for $z = 1$ and $z = -1$. This implies zeros lie on the unit circle at $w_d = 0$ and $w_d = \pi$ or $w_d = -\pi$.

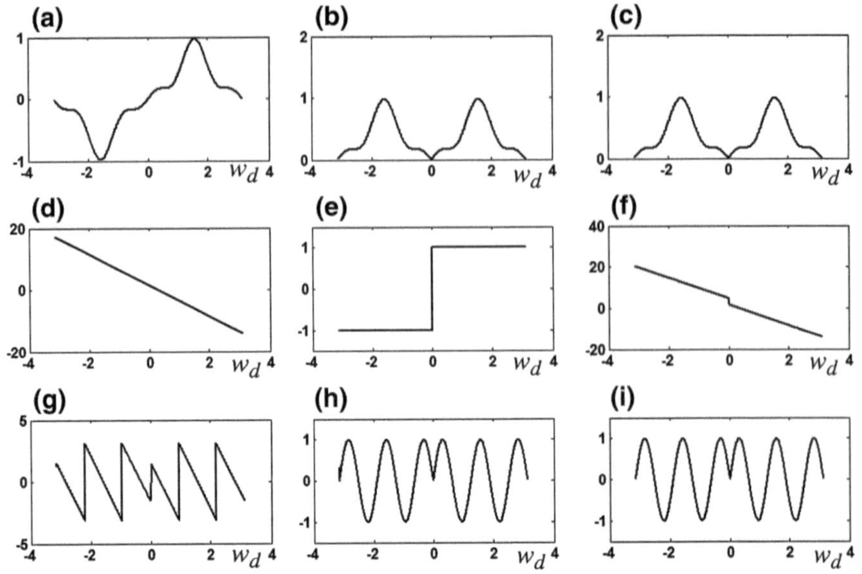

Fig. 3.2 Demonstration of Type 2 FIR filter. **a** M_2, **b** $abs(M_2)$, **c** $abs(M)$ computed for Type 2 filter, **d** P_2, **e** S = sign of M_2, **f** $PH_2 = (S-1)\frac{-\pi}{2} + P_2$, **g** P computed for Type 2 filter, **h** $cos(P)$, **i** $cos(PH_2)$

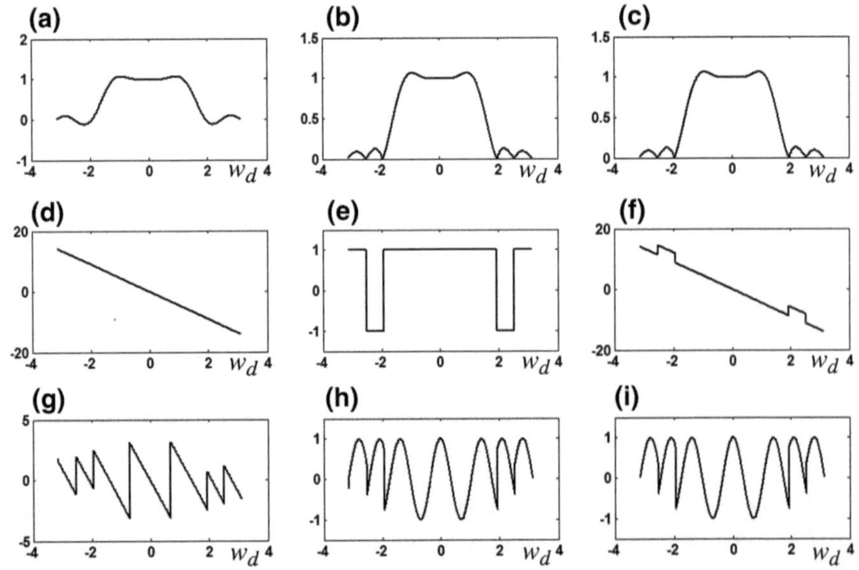

Fig. 3.3 Demonstration of Type 3 FIR filter. **a** M_3, **b** $abs(M_3)$, **c** $abs(M)$ computed for Type 3 filter, **d** P_3, **e** S = sign of M_3, **f** $PH_3 = (S-1)\frac{-\pi}{2} + P_3$, **g** P computed for Type 3 filter, **h** $cos(P)$, **i** $cos(PH_3)$

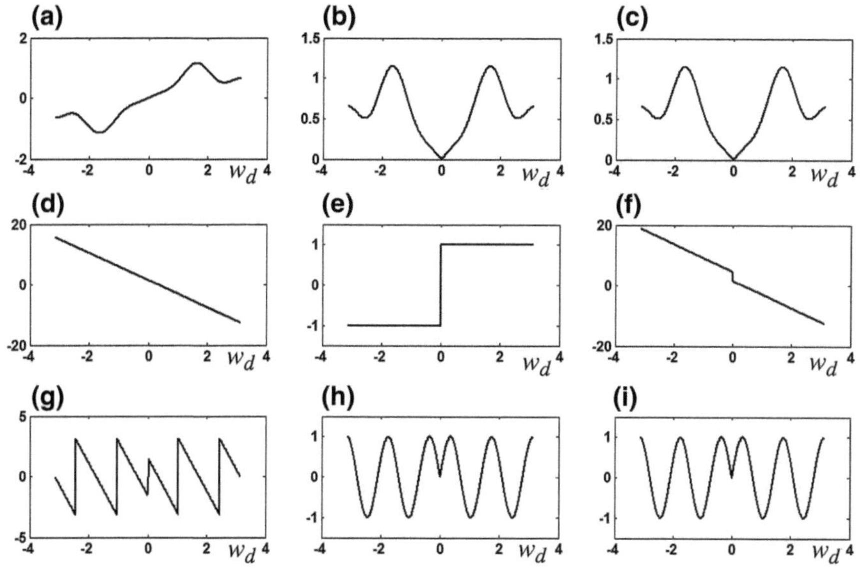

Fig. 3.4 Demonstration of Type 4 FIR filter. **a** M_4, **b** $abs(M_4)$, **c** $abs(M)$ computed for Type 4 filter, **d** P_4, **e** S = sign of M_4, **f** $PH_4 = (S-1)\frac{-\pi}{2} + P_4$, **g** P computed for Type 4 filter, **h** $cos(P)$, **i** $cos(PH_4)$

4. For Type 3 with $N = 4$, we get the polynomial as $H(z) = h(0) + h(1)z^{-1} + h(1)z^{-2} + h(0)z^{-3}$. In this case, we observe that $H(z) = 0$ for $z = -1$. This implies zero lies on the unit circle at $w_d = \pi$.
5. For Type 4 with $N = 4$, we get the polynomial as $H(z) = h(0) + h(1)z^{-1} - h(1)z^{-2} - h(0)z^{-3}$. In this case, we observe that $H(z) = 0$ for $z = 1$. This implies zero lies on the unit circle at $w_d = 0$.
6. It is observed from Fig. 3.1b, magnitude values are zero at 6 positions between $-\pi$ to π. These are due to 6 zeros present as shown in Fig. 3.5a.
7. It is observed from Fig. 3.3b, magnitude values are zero at 6 positions between $-\pi$ to π. These are due to 5 zeros present as shown in the Fig. 3.5c and one at π or $-\pi$.
8. It is observed from Fig. 3.2b, the magnitude response is Band pass in nature (These are due to presence of zeros at 0 and π) and from Fig. 3.4b, the magnitude response is High pass in nature (These are due to presence of zeros at 0). Hence Type 2 and Type 4 are not suitable for Low-pass filter design.
9. FIR filters with real coefficients (four zeros occur as the complex conjugate) that has the identical magnitude response and different phase response is illustrated in the Figs. 3.6 and 3.7.
10. If all the zeros of the transfer function $H(z)$ lie within the unit circle, it is known as minimum phase filter. If all the zeros of the transfer function $H(z)$ lie outside the unit circle, it is known as minimum phase filter.

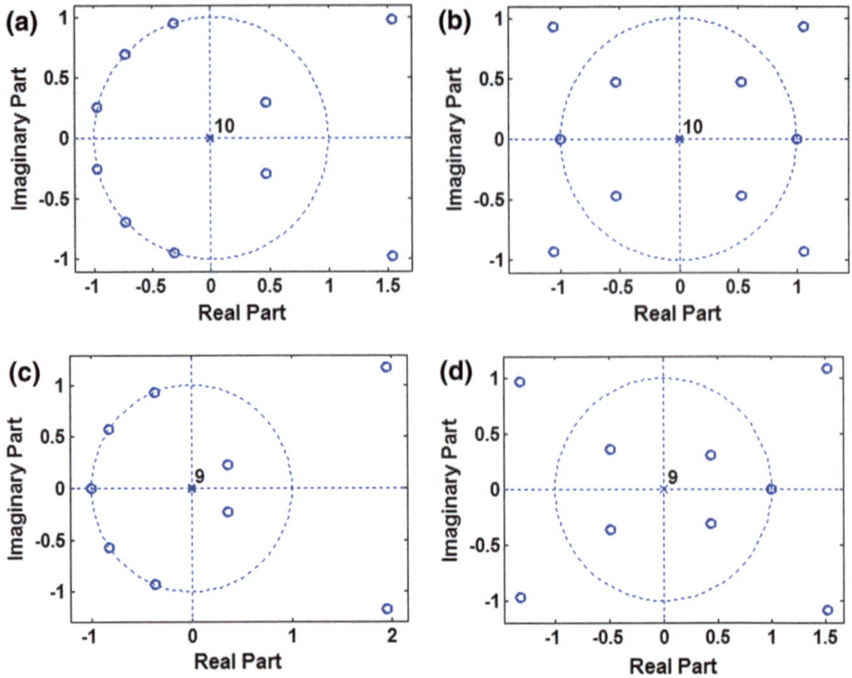

Fig. 3.5 Pole zero plot: **a** Type 1 FIR filter, **b** Type 2 FIR filter, **c** Type 3 FIR filter, **d** Type 4 FIR filter

11. From Figs. 3.6b, d and 3.7b, d, it is observed that the minimum phase FIR filter can be represented as the product of maximum phase FIR filter and the stable all pass filter (refer Fig. 3.8) (refer Sect. 3.3 for further details).

```
%FIRdemo.m
%FIR demonstration
N=11;
%%%%%%%%%%%%%%%%%%%%%%%%%%%%%%%%%%%%%%%%%%%%%%%%%%%%%%%%%%%%%%%%%%%%%%%%%%%%%%%%
%Type 1: h(n)=h(N-1-n), N->Odd

wc=pi/2;
tau=(N-1)/2;
n=0:1:N-1;
h=sin(wc*(n-tau))./(pi*(n-tau));
u=isnan(h);
[p,q]=find(u==1);
h(q)=wc/pi;
temp1=[];
temp2=[];
for w=-pi:0.001:pi;
[s,a]=firtype1(N,h,w);
temp1=[temp1 s];
temp2=[temp2 a];
end
```

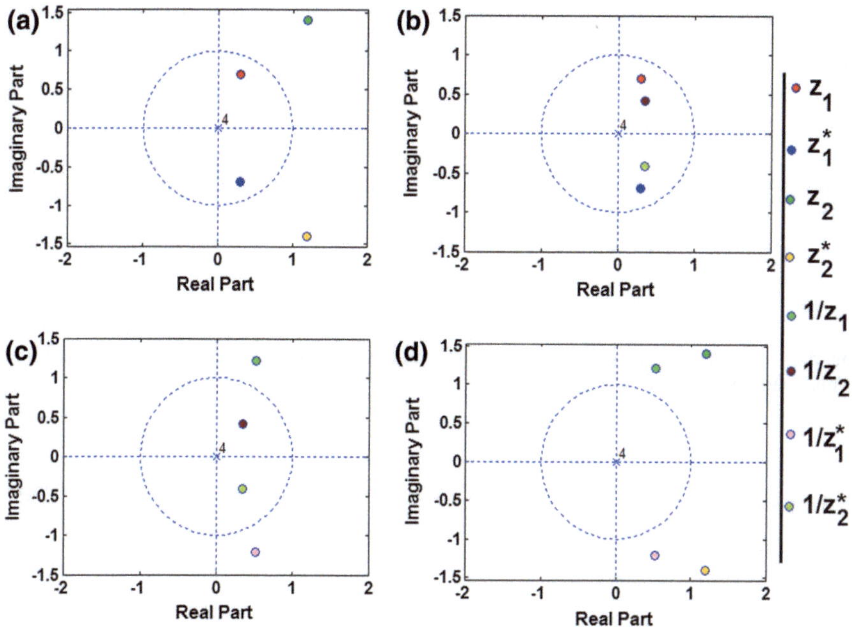

Fig. 3.6 Pole-zero plot of four filters that have identical magnitude response

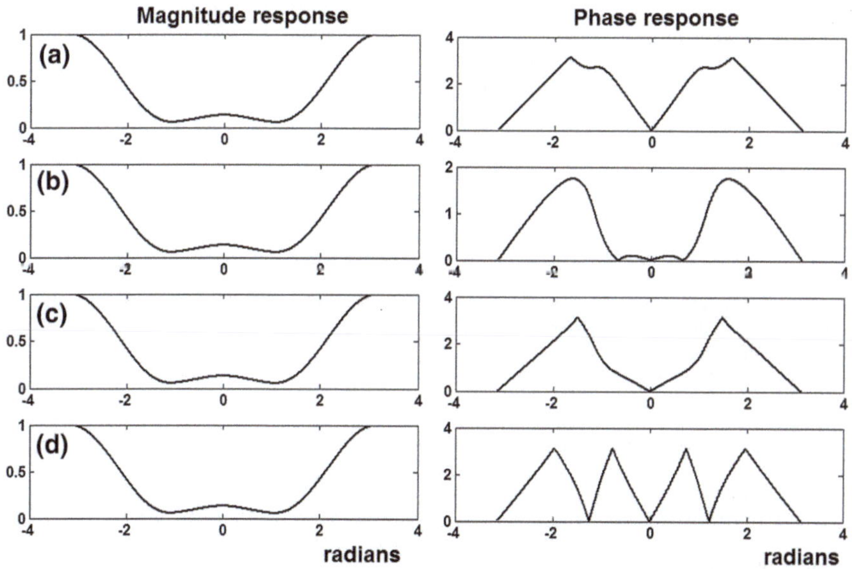

Fig. 3.7 Magnitude and phase response corresponding to the FIR filters mentioned in Fig. 3.6

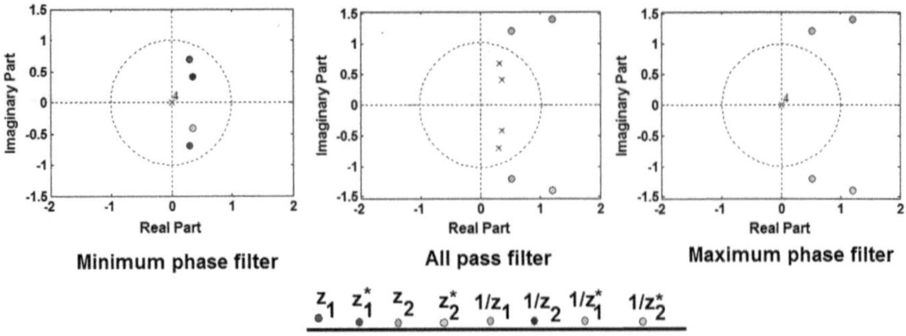

Fig. 3.8 Illustration of the pole-zero plot of the All pass filter that converts minimum phase filter to maximum phase filter

```
figure(1)

subplot(3,3,1)
plot(-pi:0.001:pi,temp1)

subplot(3,3,2)
plot(-pi:0.001:pi,abs(temp1))

m=[];
p=[];
for w=-pi:0.001:pi;
    m=[m abs(sum(h.*exp(-j*w*n)))];
    p=[p angle(sum(h.*exp(-j*w*n)))];
end

subplot(3,3,3)
plot(-pi:0.001:pi,m)

subplot(3,3,4)
plot(-pi:0.001:pi,temp2)

subplot(3,3,5)
plot(-pi:0.001:pi,sign(temp1))
temp=sign(temp1);
temp3=(temp-1)*(-pi/2);

subplot(3,3,6)
plot(-pi:0.001:pi,temp2+temp3)

subplot(3,3,7)
plot(-pi:0.001:pi,p)

subplot(3,3,8)
plot(-pi:0.001:pi,cos(p))

subplot(3,3,9)
plot(-pi:0.001:pi,cos(temp2+temp3))
figure(5)
subplot(2,2,1)
```

```
zplane(h)


%%%%%%%%%%%%%%%%%%%%%%%%%%%%%%%%%%%%%%%%%%%%%%%%%%%%%%%%%%%%%%%%%%%%%%%%
%Type 2: h(n)=-h(N-1-n), N->Odd
N=11;
wc=pi/2;
tau=(N-1)/2;
n=0:1:N-1;
h=sin(wc*(n-tau))./(pi*(n-tau));
u=isnan(h);
[p,q]=find(u==1);
h(q)=wc/pi;
h(q)=0;
for i=((N+1)/2+1):1:N
     h(i)=-1*h(i);
end
temp1=[];
temp2=[];
for w=-pi:0.001:pi;
[s,a]=firtype2(N,h,w);
temp1=[temp1 s];
temp2=[temp2 a];
end
figure(2)
subplot(3,3,1)
plot(-pi:0.001:pi,temp1)
subplot(3,3,2)
plot(-pi:0.001:pi,abs(temp1))
m=[];
p=[];
for w=-pi:0.001:pi;
    m=[m abs(sum(h.*exp(-j*w*n)))];
    p=[p angle(sum(h.*exp(-j*w*n)))];
end

subplot(3,3,3)
plot(-pi:0.001:pi,m)
subplot(3,3,4)
plot(-pi:0.001:pi,temp2)
subplot(3,3,5)
plot(-pi:0.001:pi,sign(temp1))
temp=sign(temp1);
temp3=(temp-1)*(-pi/2);
subplot(3,3,6)
plot(-pi:0.001:pi,temp2+temp3)
subplot(3,3,7)
plot(-pi:0.001:pi,p)
subplot(3,3,8)
plot(-pi:0.001:pi,cos(p))
subplot(3,3,9)
plot(-pi:0.001:pi,cos(temp2+temp3))

figure(5)
subplot(2,2,2)
zplane(h)

%%%%%%%%%%%%%%%%%%%%%%%%%%%%%%%%%%%%%%%%%%%%%%%%%%%%%%%%%%%%%%%%%%%%%%%%
```

```
%Type 3: h(n)=h(N-1-n)
N=10;
wc=pi/2;
tau=(N-1)/2;
n=0:1:N-1;
h=sin(wc*(n-tau))./(pi*(n-tau));
u=isnan(h);
[p,q]=find(u==1);
h(q)=wc/pi;
temp1=[];
temp2=[];
for w=-pi:0.001:pi;
[s,a]=firtype3(N,h,w);
temp1=[temp1 s];
temp2=[temp2 a];
end
figure(3)
subplot(3,3,1)
plot(-pi:0.001:pi,temp1)
subplot(3,3,2)
plot(-pi:0.001:pi,abs(temp1))
m=[];
p=[];
for w=-pi:0.001:pi;
    m=[m abs(sum(h.*exp(-j*w*n)))];
    p=[p angle(sum(h.*exp(-j*w*n)))];
end

subplot(3,3,3)
plot(-pi:0.001:pi,m)

subplot(3,3,4)
plot(-pi:0.001:pi,temp2)

subplot(3,3,5)
plot(-pi:0.001:pi,sign(temp1))
temp=sign(temp1);
temp3=(temp-1)*(-pi/2);

subplot(3,3,6)
plot(-pi:0.001:pi,temp2+temp3)

subplot(3,3,7)
plot(-pi:0.001:pi,p)

subplot(3,3,8)
plot(-pi:0.001:pi,cos(p))

subplot(3,3,9)
plot(-pi:0.001:pi,cos(temp2+temp3))

figure(5)
subplot(2,2,3)
zplane(h)

%Type 4: h(n)=-h(N-1-n)
N=10;
wc=pi/2;
tau=(N-1)/2;
```

```
n=0:1:N-1;
h=sin(wc*(n-tau))./(pi*(n-tau));
u=isnan(h);
[p,q]=find(u==1);
h(q)=wc/pi;
for i=((N/2)+1):1:N
     h(i)=-1*h(i);
end
temp1=[];
temp2=[];
for w=-pi:0.001:pi;
[s,a]=firtype4(N,h,w);
temp1=[temp1 s];
temp2=[temp2 a];
end
figure(4)
subplot(3,3,1)
plot(-pi:0.001:pi,temp1)
subplot(3,3,2)
plot(-pi:0.001:pi,abs(temp1))
m=[];
p=[];
for w=-pi:0.001:pi;
    m=[m abs(sum(h.*exp(-j*w*n)))];
    p=[p angle(sum(h.*exp(-j*w*n)))];
end

subplot(3,3,3)
plot(-pi:0.001:pi,m)
subplot(3,3,4)
plot(-pi:0.001:pi,temp2)
subplot(3,3,5)
plot(-pi:0.001:pi,sign(temp1))
temp=sign(temp1);
temp3=(temp-1)*(-pi/2);
subplot(3,3,6)
plot(-pi:0.001:pi,temp2+temp3)
subplot(3,3,7)
plot(-pi:0.001:pi,p)
subplot(3,3,8)
plot(-pi:0.001:pi,cos(p))
subplot(3,3,9)
plot(-pi:0.001:pi,cos(temp2+temp3))

figure(5)
subplot(2,2,4)
zplane(h)


%firtype1.m
function [s,a]=firtype1(N,h,w)
s=0;
tau=(N-1)/2;
for m=1:1:tau
    s=s+2*h(m)*cos((tau-m+1)*w);
end
s=s+h(tau+1);
a=-tau*w;
```

```
firtype2.m
function [s,a]=firtype2(N,h,w)
s=0;
tau=(N-1)/2;
for m=1:1:tau
    s=s+2*h(m)*sin((tau-m+1)*w);
end
s=s+h(tau+1);
a=-tau*w+(pi/2);

firtype3.m
function [s,a]=firtype3(N,h,w)
s=0;
tau=(N-1)/2;
for m=1:1:(N/2)
    s=s+2*h(m)*cos((tau-m+1)*w);
end
a=-tau*w;

firtype4.m
function [s,a]=firtype4(N,h,w)
s=0;
tau=(N-1)/2;
for m=1:1:(N/2)
    s=s+2*h(m)*sin((tau-m+1)*w);
end
a=-tau*w+(pi/2);


%minmaxFIR.m
%Demonstration of minimum phase FIR filter
%and maximum phase FIR filter
z1=0.3+0.7j;
z2=1.2+1.4j;
z3=z1';
z4=z2';
%h=(z-z1)(z-z2)(z-z1')(z-z2')
h1=conv([1 -z1],[1 -z2]);
h2=conv(h1,[1 -z1'] );
h3=conv(h2,[1 -z2'] );
figure(1)
subplot(2,2,1)
zplane(h3,1)
h=h3;
mag=[];
phase=[];
n=0:1:length(h3)-1;
for w=-pi:0.001:pi;
    mag=[mag abs(sum(h.*exp(-j*w*n)))];
    phase=[phase angle(sum(h.*exp(-j*w*n)))];
end
figure(2)
subplot(4,2,1)
plot(-pi:0.001:pi,abs(mag)/max(abs(mag)))
subplot(4,2,2)
plot(-pi:0.001:pi,abs(phase))
%%%%%%%%%%%%%%%%%%%%%%%%%%%%%%%%%%%%%%%
%h=(z-z1)(z-z1')(z-1/z2')(z-(1/z2')')
h1=conv([1 -z1],[1 -1/z2]);
```

```
h2=conv(h1,[1 -z1'] );
h3=conv(h2,[1 (-1/z2)'] );
h=h3;
figure(1)
subplot(2,2,2)
zplane(h3,1)
mag=[];
phase=[];
for w=-pi:0.001:pi;
    mag=[mag abs(sum(h.*exp(-j*w*n)))];
    phase=[phase angle(sum(h.*exp(-j*w*n)))];
end
figure(2)
subplot(4,2,3)
plot(-pi:0.001:pi,abs(mag)/max(abs(mag)))
subplot(4,2,4)
plot(-pi:0.001:pi,abs(phase))

%%%%%%%%%%%%%%%%%%%%%%%%%%%%%%%%%%%
%h=(z-1/z1)(z-(1/z1')')(z-1/z2')(z-(1/z2')')
h1=conv([1 -1/z1],[1 (-1/z1)']);
h2=conv(h1,[1 -1/z2'] );
h3=conv(h2,[1 (-1/z2')'] );
h=h3;
figure(1)
subplot(2,2,3)
zplane(h3,1)
mag=[];
phase=[];
for w=-pi:0.001:pi;
    mag=[mag abs(sum(h.*exp(-j*w*n)))];
    phase=[phase angle(sum(h.*exp(-j*w*n)))];
end
figure(2)
subplot(4,2,5)
plot(-pi:0.001:pi,abs(mag)/max(abs(mag)))
subplot(4,2,6)
plot(-pi:0.001:pi,abs(phase))

%%%%%%%%%%%%%%%%%%%%%%%%%%%%%%%%%%%
%h=(z-1/z1')(z-(1/z1')')(z-z2)(z-z2')
h1=conv([1 -1/z1'],[1 (1/-z1')']);
h2=conv(h1,[1 -z2] );
h3=conv(h2,[1 -z2'] );
h=h3;
figure(1)
subplot(2,2,4)
zplane(h3,1)
mag=[];
phase=[];
for w=-pi:0.001:pi;
    mag=[mag abs(sum(h.*exp(-j*w*n)))];
    phase=[phase angle(sum(h.*exp(-j*w*n)))];
end
figure(2)
subplot(4,2,7)
plot(-pi:0.001:pi,abs(mag)/max(abs(mag)))
subplot(4,2,8)
plot(-pi:0.001:pi,abs(phase))
```

3.2 Design of Linear Phase FIR Filter-Windowing Technique

3.2.1 *Design of Low-Pass Filter*

Non-causal IIR filter is obtained by computing inverse discrete Fourier transformation (IDFT) of the transfer function $H_d(e^{jw_d}) = 1$, for $-w_c \leq w_d \leq w_c, 0$, otherwise. $h_d(n)$ for $n = \cdots, -2, -1, 0, 1, 2, \cdots$ is computed as follows.

$$h_d(n) = \frac{1}{2\pi}\int_{-\pi}^{\pi} H_d(e^{jw_d})e^{jw_d n}dw_d \tag{3.9}$$

$$= \frac{1}{2\pi}\int_{-w_c}^{w_c} e^{jw_d n}dw_d = \frac{1}{2\pi}(\frac{e^{jw_c n}}{jn} - \frac{e^{-jw_c n}}{jn}) \tag{3.10}$$

$$= \frac{1}{2\pi}\frac{2jsin(w_c n)}{jn} = \frac{1}{2\pi}\frac{2sin(w_c n)}{n} = \frac{sin(w_c n)}{\pi n} \tag{3.11}$$

FIR filter impulse response $(h(n))$ is obtained by truncating the impulse response of the IIR filter ($h_d(n)$ for $-10 \leq n \leq 10$). This is obtained by multiplying IIR filter with the rectangular window $w(n) = 1$ for $-10 \leq n \leq 10$ and 0, otherwise. The truncated IIR filter impulse response and the corresponding magnitude and phase response is illustrated in Fig. 3.9.

$$h(n) = h_d(n)w(n) \tag{3.12}$$

$$\Rightarrow H(e^{jw_d}) = H_d(e^{jw_d}) * W(e^{jw_d}) \tag{3.13}$$

Figure 3.10 demonstrates that the ripples are introduced in the transfer function $H(e^{jw_d})$ due to the convolution of $H_d(e^{jw_d})$ and $W(e^{jw_d})$. The graph is plotted for $-\pi \leq w_d \leq \pi$ and the amplitude is normalized to 1. The impulse response $h(n)$ is made causal $(hc(n))$ by shifting the impulse response sequence toward right, such that is zero for $n < 0$. From (3.11), for $N = 21$, causal impulse response $hc(n)$ is given as the

$$hc(n) = \frac{sin(w_c(n - \frac{(N-1)}{2}))}{\pi(n - \frac{(N-1)}{2})} \tag{3.14}$$

This forms the Type 1 FIR filter as described in the Sect. 3.1. $h(n)$ can be viewed as sampling $h(t) = \frac{sin(w_c t)}{w_c t}$ for the duration $\frac{-(N-1)T_s}{2} \leq t \leq \frac{(N-1)T_s}{2}$, with N as odd. It is made causal by shifting the impulse response toward right. The method of obtaining Type 1 and Type 3 FIR filter by sampling the impulse response of the truncated FIR filter is illustrated in Fig. 3.11. The corresponding magnitude and phase response is given in Fig. 3.12. Similarly, the method of obtaining Type 2,

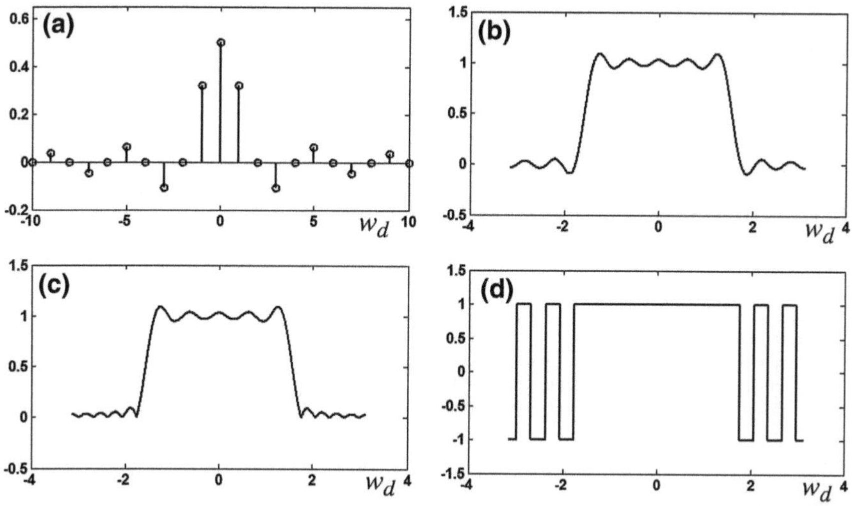

Fig. 3.9 **a** Impulse response of the truncated filter. **b** Transfer function of the truncated filter. **c** Magnitude response of the truncated filter. **d** Cosine of the phase response of the truncated filter. (It is noted that the value is 1, if the sign of the transfer function is positive and the value is -1 corresponding to the case when the transfer functional value is negative.)

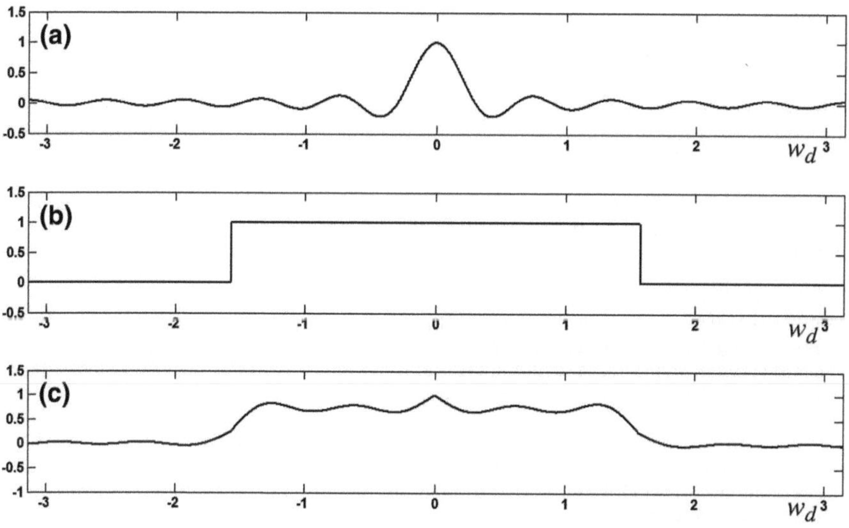

Fig. 3.10 **a** Transfer function of the rectangular pulse $W(e^{jw_d})$. **b** Transfer function of the ideal low pass filter $H_d(e^{jw_d})$. **c** Magnitude response of the truncated filter $H(e^{jw_d})$

Type 4 FIR filter by modifying the samples obtained from truncated $h(t)$, satisfying the $h(n) = -h(N - n - 1)$ is illustrated in Fig. 3.13. The corresponding magnitude and phase response is given in Fig. 3.14. Note that the magnitude response of Type 2 and Type 4 filter is of High pass in nature.

```
%effectoftruncation.m
%Demonstration of magnitude response of the truncated filter
n=-10:1:10;
wc=pi/2;
h=sin(wc*n)./(pi*n);
h(11)=wc/pi;
figure
subplot(2,2,1)
stem(n,h)
title('Impulse response of the truncated filter')
m1=[];
m2=[];
p=[];
for w=-pi:0.001:pi;
    m1=[m1 sum(h.*exp(-j*w*n))];
    m2=[m2 abs(sum(h.*exp(-j*w*n)))];
    p=[p angle(sum(h.*exp(-j*w*n)))];
end
temp=sign(m1);
w=-pi:0.001:pi
subplot(2,2,2)
plot(w,real(m1))
title('Transfer function of the truncated filter')
subplot(2,2,3)
plot(w,real(m2))
title('Magnitude response of the truncated filter')
subplot(2,2,4)
plot(w,cos(p))
title('Cosine of the phase response of the truncated filter')


energy=sum(h.^2);
rw=sqrt(energy/21);
rw=ones(1,21)*rw;
tfrw=[];
for w=-2*pi:0.001:2*pi;
    tfrw=[tfrw sum(rw.*exp(-j*w*n))];
end

s1=-2*pi:0.001:(-3*pi/2);
s2=s1(length(s1))+0.001:0.001:(-pi/2);
s3=s2(length(s2))+0.001:0.001:pi/2;
s4=s3(length(s3))+0.001:0.001:(3*pi/2);
s5=s4(length(s4))+0.001:0.001:2*pi;

Hdw=[ones(1,length(s1)) zeros(1,length(s2)) ones(1,length(s3)) ...
zeros(1,length(s4)) ones(1,length(s5))];
res=conv(real(tfrw),Hdw);
res=(res/max(res))*21;
figure
subplot(3,1,1)
L1=length(-2*pi:0.001:2*pi); L2=length(-2*pi:0.001:-pi-0.001);
plot(-pi:0.001:pi,
real(tfrw(L2+1:1:L1-L2-1)/max(tfrw(L2+1:1:L1-L2-1))));
title('Transfer function of the rectangular pulse W(e^{jw_{d}})')
subplot(3,1,2) plot(-pi:0.001:pi, Hdw(L2+1:1:L1-L2-1));
title('Transfer function of the ideal low-pass filter H_{d}(e^{jw_{d}})')
```

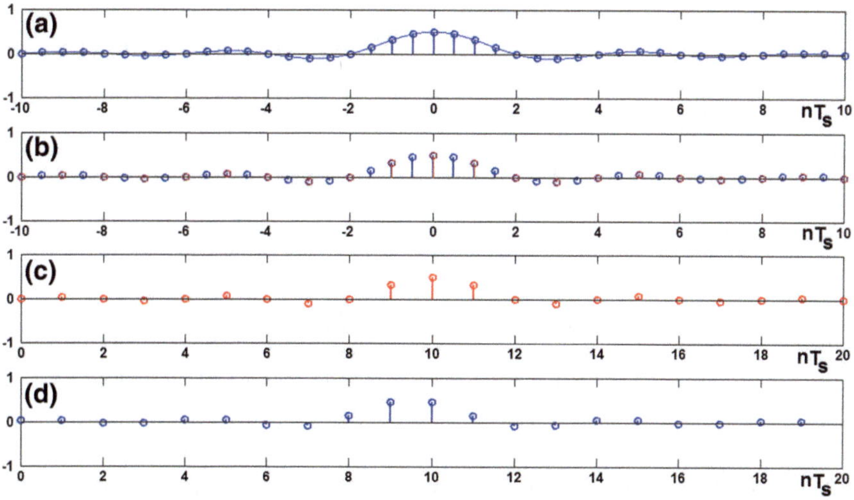

Fig. 3.11 **a** Impulse response of the truncated Low-pass filter (with the specific cutoff frequency) with the samples shown corresponding to half the sampling time. **b** Illustration of samples corresponding to the integer multiples of the sampling time (*red*) and integer multiples of the half the sampling time (*blue*). **c** *Red samples* are collected to obtain the impulse response of the type 1 FIR filter. **d** *Blue samples* are collected to obtain the impulse response of the corresponding type 3 FIR filter

```
subplot(3,1,3)
L3=length(-4*pi:0.001:4*pi);
L4=length(-4*pi:0.001:-pi-0.001);
plot(-pi:0.001:pi, res(L4+1:1:L3-L4-1)/max(res(L4+1:1:L3-L4-1)));
title('Magnitude response of the truncated filter H(e^{jw_{d}}))
```

```
%truncimpLPF2FIR1FIR3.m
%Illustration of obtaining the filter ...
%coefficients for FIR 1 and FIR 3 filter from ...
%the truncated impulse response of the ideal Low-pass filter
t=-10:0.01:10;
wc=pi/2;
ht=sin(wc*t)./(pi*t);
ht(1001)=wc/pi;
figure
subplot(4,1,1)
plot(t,ht)
hold on
n1=-10:0.5:10;
wc=pi/2;
h=sin(wc*n1)./(pi*n1);
h(21)=wc/pi;
hold on
stem(n1,h)
subplot(4,1,2)
stem(n1,h)
hold on
```

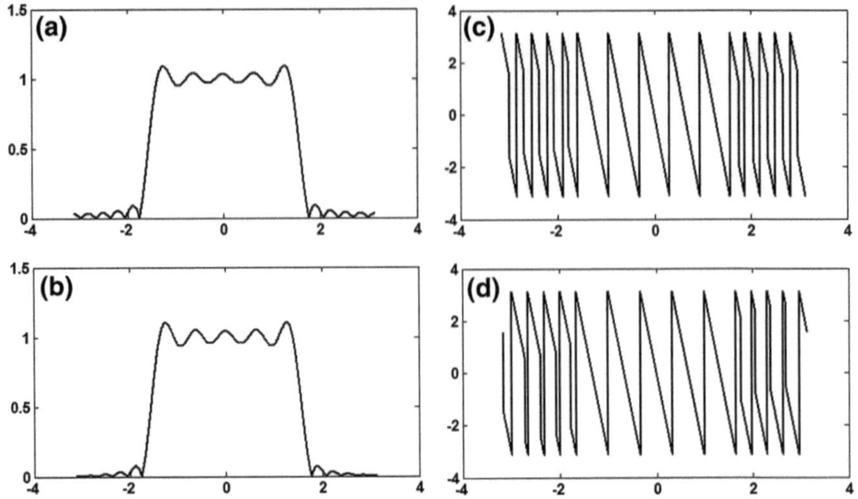

Fig. 3.12 Magnitude and phase response of the Impulse response mentioned in Fig. 3.11c, d. Subplot **a**, **b** corresponds to Type 1 FIR filter. Subplot **c**, **d** corresponds to Type 3 FIR filter

```
n2=n1(1:2:41);
stem(n2,h(1:2:41),'r')
subplot(4,1,3)
stem(0:1:20,h(1:2:41),'r')
subplot(4,1,4)
stem(0:1:19,h(2:2:41))
h1fir=h(1:2:41);
h3fir=h(2:2:41);

n=0:1:length(h1fir)-1;
mag1=[];
phase1=[];
for w=-pi:0.001:pi;
    mag1=[mag1 abs(sum(h1fir.*exp(-j*w*n)))];
    phase1=[phase1 angle(sum(h1fir.*exp(-j*w*n)))];
end

n=0:1:length(h3fir)-1;
mag2=[];
phase2=[];
for w=-pi:0.001:pi;
    mag2=[mag2 abs(sum(h3fir.*exp(-j*w*n)))];
    phase2=[phase2 angle(sum(h3fir.*exp(-j*w*n)))];
end
figure
subplot(2,2,1)
plot(-pi:0.001:pi,mag1)
subplot(2,2,2)
plot(-pi:0.001:pi,phase1)
subplot(2,2,3)
plot(-pi:0.001:pi,mag2)
subplot(2,2,4)
plot(-pi:0.001:pi,phase2)
```

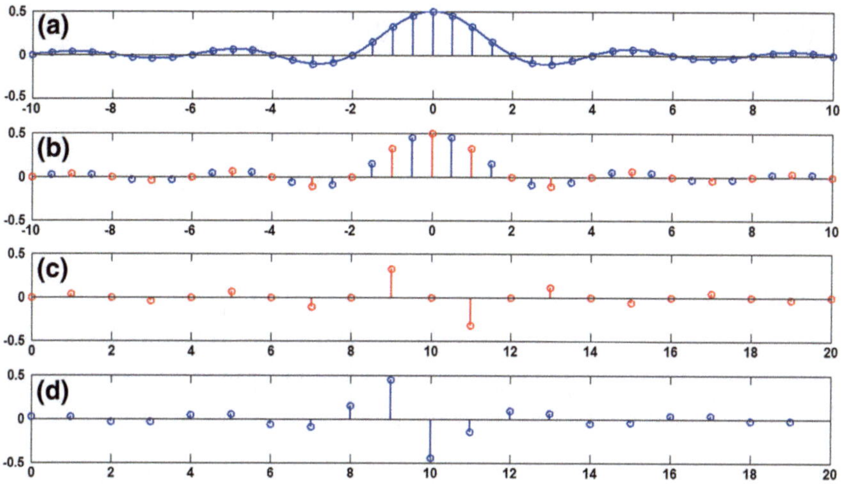

Fig. 3.13 **a** Impulse response of the truncated Low-pass filter (with the specific cutoff frequency) with the samples shown corresponding to half the sampling time. **b** Illustration of samples corresponding to the integer multiples of the sampling time (*red*) and integer multiples of the half the sampling time (*blue*). **c** *Red samples* are collected to obtain the impulse response of the type 2 FIR filter. **d** *Blue samples* are collected to obtain the impulse response of the corresponding type 4 FIR filter

```
%truncimpLPF2FIR2FIR4.m
%Illustration of obtaining the filter ...
%coefficients for FIR 2 and FIR 4 filter from ...
%the truncated impulse response of the ideal Low-pass filter
t=-10:0.01:10;
wc=pi/2;
ht=sin(wc*t)./(pi*t);
ht(1001)=wc/pi;
figure
subplot(4,1,1)
plot(t,ht)
hold on
n1=-10:0.5:10;
wc=pi/2;
h=sin(wc*n1)./(pi*n1);
h(21)=wc/pi;
hold on
stem(n1,h)
subplot(4,1,2)
stem(n1,h)
hold on
n2=n1(1:2:41);
stem(n2,h(1:2:41),'r')
%Type 2 filter
fir2h=h(1:2:41);
fir2h(11)=0;
for i=12:1:21
    fir2h(i)=-fir2h(21-i+1);
end
%Type 4 filter
fir4h=h(2:2:41);
```

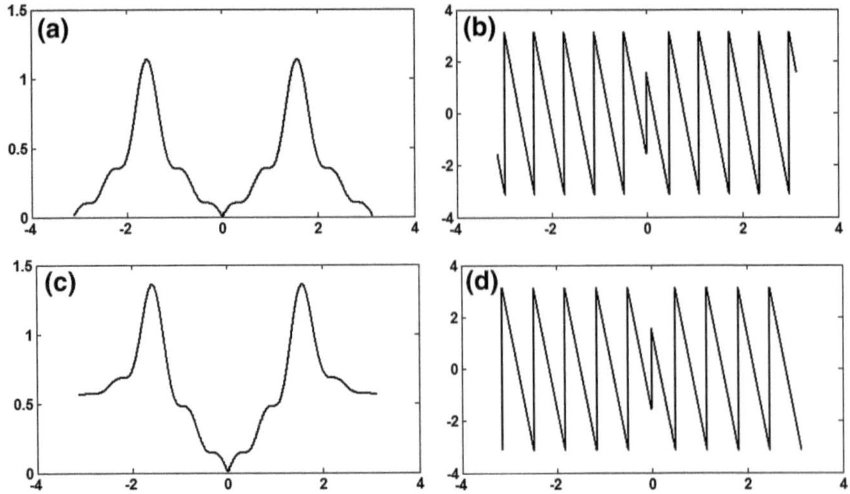

Fig. 3.14 Magnitude and phase response of the Impulse response mentioned in Fig. 3.13c, d. Subplot **a**, **b** corresponds to Type 2 FIR filter. Subplot **c**, **d** corresponds to Type 4 FIR filter

```
for i=11:1:20
    fir4h(i)=-fir4h(20-i+1);
end
subplot(4,1,3)
stem(0:1:20,fir2h,'r')
subplot(4,1,4)
stem(0:1:19,fir4h,'b')

n=0:1:length(fir2h)-1;
mag1=[];
phase1=[];
for w=-pi:0.001:pi;
    mag1=[mag1 abs(sum(fir2h.*exp(-j*w*n)))];
    phase1=[phase1 angle(sum(fir2h.*exp(-j*w*n)))];
end
n=0:1:length(fir4h)-1;
mag2=[];
phase2=[];
for w=-pi:0.001:pi;
    mag2=[mag2 abs(sum(fir4h.*exp(-j*w*n)))];
    phase2=[phase2 angle(sum(fir4h.*exp(-j*w*n)))];
end
figure
subplot(2,2,1)
plot(-pi:0.001:pi,mag1)
subplot(2,2,2)
plot(-pi:0.001:pi,phase1)
subplot(2,2,3)
plot(-pi:0.001:pi,mag2)
subplot(2,2,4)
plot(-pi:0.001:pi,phase2)
```

3.2.2 Design of High-Pass Filter

The ideal magnitude response of the High-pass filter is obtained by shifting the plot of the magnitude response of the ideal low-pass filter (refer Fig. 3.9b, say $H_{LP}(e^{jw_d})$) toward right by π. This is obtained by replacing w_d by $w_d - \pi$ in $H_{LP}(e^{jw_d})$ of the low-pass filter as follows.

$$H_{LP}(e^{jw_d}) = h(0) + h(1)e^{-jw_d} + h(2)e^{-j2w_d} + \cdots + h(N-1)e^{-j(N-1)w_d}$$
$$H_{HP}(e^{jw_d}) = H_{LP}(e^{j(w_d-\pi)})$$
$$= h(0) - h(1)e^{-jw_d} + h(2)e^{-j2w_d} + \cdots - h(N-1)e^{-j(N-1)w_d}$$
$$\Rightarrow h_{HP}(n) = (-1)^n h(n)$$

where, $h(n) = \frac{sin(w_c(n-\frac{(N-1)}{2}))}{\pi(n-\frac{(N-1)}{2})}$ and it satisfies $h_{HP}(n) = -h_{HP}(N-n-1)$. This can also be obtained by subtracting the impulse response of two low-pass filter $\frac{sin(\pi(n-\frac{(N-1)}{2}))}{\pi(n-\frac{(N-1)}{2})}$ (with cutoff frequency π radians) and $\frac{sin(w_c(n-\frac{(N-1)}{2}))}{\pi(n-\frac{(N-1)}{2})}$ (with cutoff frequency w_c radians). Hence the impulse response of the high-pass filter is given as the following.

$$\left(\frac{sin(\pi(n-\frac{(N-1)}{2}))}{\pi(n-\frac{(N-1)}{2})}\right) - \left(\frac{sin(w_c(n-\frac{(N-1)}{2}))}{\pi(n-\frac{(N-1)}{2})}\right) \tag{3.15}$$

As the phase response of the individual low-pass filters are linear in nature, the phase response of the high-pass filter constructed using the low-pass filter is also linear. Refer Fig. 3.16 to see the magnitude and phase response of HPF with cutoff frequency $\frac{\pi}{2}$ obtained using the Type 1 and Type 3 LPF. It is observed from the figure that number of zero crossings in the magnitude response is larger for the Type 1 LPF based design.

3.2.3 Design of Band Pass and Band Reject Filter

1. Band pass filter with pass band frequencies ranging from w_1 to w_2 (with $w_1 \leq w_2$) is obtained as the cascade connection of two filters (refer Fig. 3.15b), i.e., Low-pass filter with cutoff w_2 radians, followed by High pass filter with cutoff frequency w_1 radians.
2. Similarly, Band reject filter with the stop band frequencies ranging from $w_1 to w_2$ (with $w_1 \leq w_2$) is obtained as the parallel realization of two filters (refer Fig. 3.15a), i.e., Low-pass filter with cutoff w_1 radians, and the High-pass filter with cutoff frequency w_2 radians.

3. Figure 3.17a illustrates the typical magnitude response of the band pass filter obtained as the cascade of Type 1 LPF and Type 1 HPF. Figure 3.17b illustrates the typical magnitude response of the band pass filter obtained as the cascade of Type 3 LPF and Type 3 HPF. Figure 3.17c illustrates the typical magnitude response of the band pass filter obtained as the cascade of Type 3 LPF and Type 1 HPF. Finally Fig. 3.17d illustrates the typical magnitude response of the band pass filter obtained as the cascade of Type 1 LPF and Type 3 HPF. It is observed that the number of zero crossings in the pass band of the band pass filter is larger when the identical types are used for LPF and HPF.
4. Figure 3.18a illustrates the typical magnitude response of the band reject filter obtained as the parallel realization of Type 1 LPF and Type 1 HPF. Figure 3.18b illustrates the typical magnitude response of the band reject filter obtained as the parallel realization of Type 3 LPF and Type 3 HPF. Figure 3.18d illustrates the typical magnitude response of the band reject filter obtained as the parallel realization of Type 3 LPF and Type 1 HPF. Finally Fig. 3.18c illustrates the typical magnitude response of the band reject filter obtained as the parallel realization of Type 1 LPF and Type 3 HPF. It is observed that the number of zero crossings in the pass band of the band pass filter is larger when the identical types are used for LPF and HPF.
5. The impulse response of the Band reject filter with stop band range from w_{c1} to w_{c2} is given as

$$h_{Bandreject}(n) = h_1(n) + h_2(n)$$

$$h_1(n) = \frac{sin(w_{c1}(n - \frac{(N-1)}{2}))}{\pi(n - \frac{(N-1)}{2})}$$

$$h_2(n) = \frac{sin(\pi(n - \frac{(N-1)}{2}))}{\pi(n - \frac{(N-1)}{2})} - \frac{sin(w_{c2}(n - \frac{(N-1)}{2}))}{\pi(n - \frac{(N-1)}{2})}$$

6. The impulse response of the Band pass filter with pass range w_{c1} to w_{c2}, is given as

$$h_{Bandpass} = h_1(n) * h_2(n)$$

$$h_1(n) = \frac{sin(\pi(n - \frac{(N-1)}{2}))}{\pi(n - \frac{(N-1)}{2})} - \frac{sin(w_{c2}(n - \frac{(N-1)}{2}))}{\pi(n - \frac{(N-1)}{2})}$$

$$h_2(n) = \frac{sin(w_{c1}(n - \frac{(N-1)}{2}))}{\pi(n - \frac{(N-1)}{2})}$$

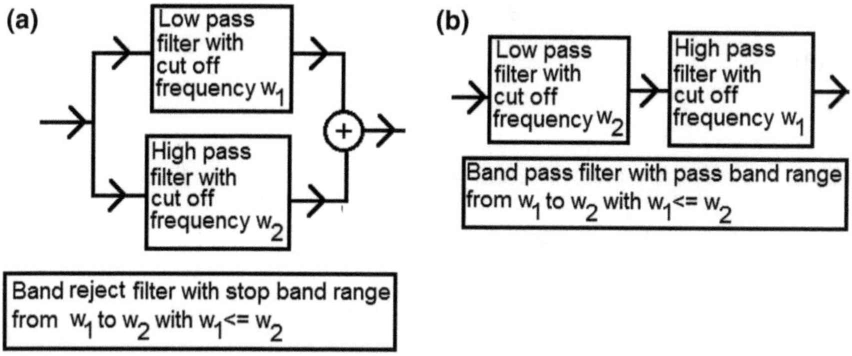

Fig. 3.15 **a** Block diagram to realize band reject filter. **b** Block diagram to realize band pass filter

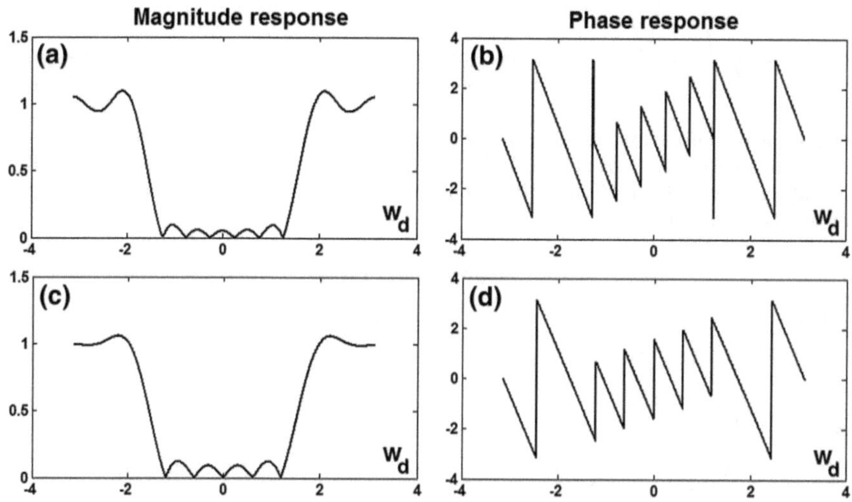

Fig. 3.16 **a** Magnitude response of the HPF using Type 1 LPF. **b** Phase response of the HPF using Type 1 LPF. **c** Magnitude response of the HPF using Type 3 LPF. **d** Phase response of the HPF using Type 3 LPF

```
%HPFtype1.m
%HPF: Designed using Type 1 (h(n)=h(N-1-n), N->Odd LPF)
N=11;
wc=pi/2;
tau=(N-1)/2;
n=0:1:N-1;
h=sin(wc*(n-tau))./(pi*(n-tau));
u=isnan(h);
[p,q]=find(u==1);
h(q)=wc/pi;
h=((-1*ones(1,length(h))).^n).*h;
m=[];
p=[];
for w=-pi:0.001:pi;
```

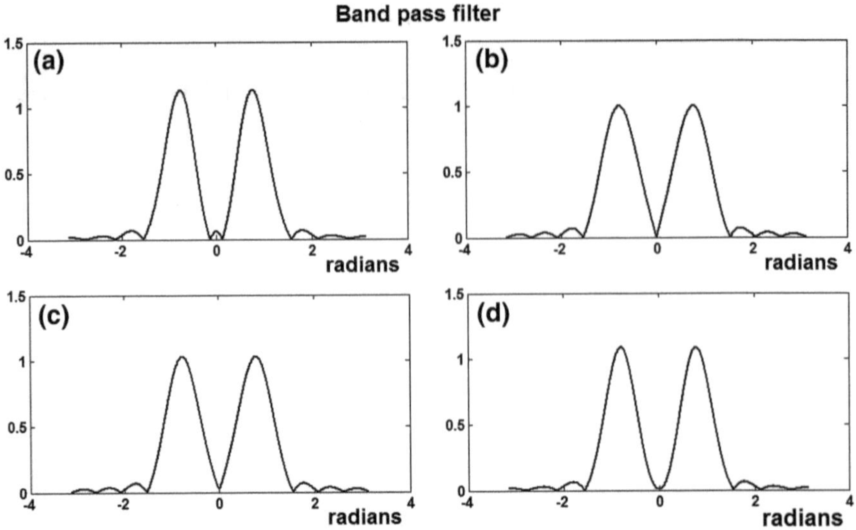

Fig. 3.17 **a** Magnitude response of the BPF (Cascade of Type 1 LPF and Type 1 HPF). **b** Magnitude response of the BPF (Cascade of Type 3 LPF and Type 3 HPF). **c** Magnitude response of the BPF (Cascade of Type 3 LPF and Type 1 HPF). **d** Magnitude response of the BPF (Cascade of Type 1 LPF and Type 3 HPF)

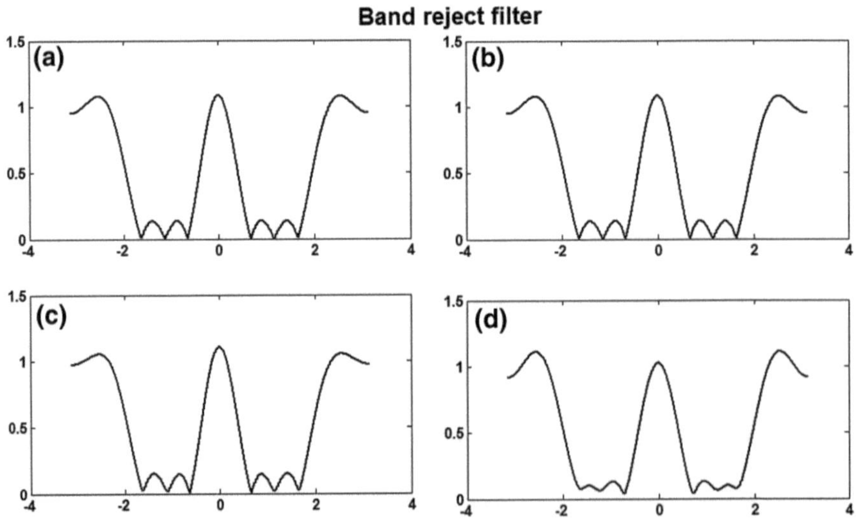

Fig. 3.18 **a** Magnitude response of the BRF (Cascade of Type 1 LPF and Type 1 HPF), **b** magnitude response of the BRF (Cascade of Type 3 LPF and Type 3 HPF), **c** magnitude response of the BRF (Cascade of Type 1 LPF and Type 3 HPF), **d** magnitude response of the BPF (Cascade of Type 3 LPF and Type 1 HPF)

```
    m=[m abs(sum(h.*exp(-j*w*n)))];
    p=[p angle(sum(h.*exp(-j*w*n)))];
end
figure
subplot(2,2,1)
plot(-pi:0.001:pi,m)
subplot(2,2,2)
plot(-pi:0.001:pi,p)
```

```
%HPFtype3.m
%HPF: Designed using Type 3 LPF (h(n)=h(N-1-n), N->Even)
N=10;
wc=pi/2;
tau=(N-1)/2;
n=0:1:N-1;
h=sin(wc*(n-tau))./(pi*(n-tau));
u=isnan(h);
[p,q]=find(u==1);
h(q)=wc/pi;
h=((-1*ones(1,length(h))).^n).*h;
m=[];
p=[];
for w=-pi:0.001:pi;
    m=[m abs(sum(h.*exp(-j*w*n)))];
    p=[p angle(sum(h.*exp(-j*w*n)))];
end
subplot(2,2,3)
plot(-pi:0.001:pi,m)
subplot(2,2,4)
plot(-pi:0.001:pi,p)
```

```
%BPFT1LPFT1HPF.m
% BPF: Designed as the cascade of Type 1 LPF (h(n)=h(N-1-n),
%N->Odd) and  HPF (obtained using Type 1 LPF)with
%pass band between pi/8 and 3*pi/8
N=11;
wc1=pi-pi/8;
wc2=3*pi/8;
tau=(N-1)/2;
n=0:1:N-1;
h1=sin(wc2*(n-tau))./(pi*(n-tau));
u=isnan(h1);
[p,q]=find(u==1);
h1(q)=wc2/pi;
n=0:1:N-1;
h2=sin(wc1*(n-tau))./(pi*(n-tau));
u=isnan(h2);
[p,q]=find(u==1);
h2(q)=wc1/pi;
h2=((-1*ones(1,length(h2))).^n).*h2;
h=conv(h1,h2);
n=0:1:length(h)-1;
m=[];
for w=-pi:0.001:pi;
    m=[m abs(sum(h.*exp(-j*w*n)))];
end
figure
```

```
subplot(2,2,1)
plot(-pi:0.001:pi,m)

%BPFT3LPFT3HPF.m
%BPF: Designed as the cascade realization of Type 3 LPF
%(h(n)=h(N-1-n), N->Even) and HPF (obtained using Type 3 LPF)
%with pass band between pi/8 and 3*pi/8
N=10;
wc1=pi-pi/8;
wc2=3*pi/8;
tau=(N-1)/2;
n=0:1:N-1;
h1=sin(wc2*(n-tau))./(pi*(n-tau));
u=isnan(h1);
[p,q]=find(u==1);
h1(q)=wc2/pi;

n=0:1:N-1;
h2=sin(wc1*(n-tau))./(pi*(n-tau));
u=isnan(h2);
[p,q]=find(u==1);
h2(q)=wc1/pi;
h2=((-1*ones(1,length(h2))).^n).*h2;
h=conv(h1,h2);
m=[];
n=0:1:length(h)-1;
for w=-pi:0.001:pi;
    m=[m abs(sum(h.*exp(-j*w*n)))];
end
subplot(2,2,2)
plot(-pi:0.001:pi,m)

%BPFT3LPFT1HPF.m
%BPF: Designed as the cascade realization of Type 3 LPF
%(h(n)=h(N-1-n), N->Even) and HPF (obtained using
%Type 1 LPF (h(n)=h(N-1-n), N->Odd) )
%with pass band between pi/8 and 3*pi/8
N=10;
wc1=pi-pi/8;
wc2=3*pi/8;
tau=(N-1)/2;
n=0:1:N-1;
h1=sin(wc2*(n-tau))./(pi*(n-tau));
u=isnan(h1);
[p,q]=find(u==1);
h1(q)=wc2/pi;
N=11;
n=0:1:N-1;
h2=sin(wc1*(n-tau))./(pi*(n-tau));
u=isnan(h2);
[p,q]=find(u==1);
h2(q)=wc1/pi;
h2=((-1*ones(1,length(h2))).^n).*h2;
h=conv(h1,h2);
m=[];
n=0:1:length(h)-1;
for w=-pi:0.001:pi;
    m=[m abs(sum(h.*exp(-j*w*n)))];
```

```
end
subplot(2,2,3)
plot(-pi:0.001:pi,m)


%BPFT1LPFT3HPF.m
%BPF: Designed as the cascade realization of Type 1 LPF (h(n)=h(N-1-n),
N->Odd) and
% HPF (obtained using Type 3 LPF (h(n)=h(N-1-n), N->Even) )
%with pass band between pi/8 and 3*pi/8
N=11;
wc1=pi-pi/8;
wc2=3*pi/8;
tau=(N-1)/2;
n=0:1:N-1;
h1=sin(wc2*(n-tau))./(pi*(n-tau));
u=isnan(h1);
[p,q]=find(u==1);
h1(q)=wc2/pi;
N=10;
n=0:1:N-1;
h2=sin(wc1*(n-tau))./(pi*(n-tau));
u=isnan(h2);
[p,q]=find(u==1);
h2(q)=wc1/pi;
h2=((-1*ones(1,length(h2))).^n).*h2;
h=conv(h1,h2);
m=[];
n=0:1:length(h)-1;
for w=-pi:0.001:pi;
    m=[m abs(sum(h.*exp(-j*w*n)))];
end
subplot(2,2,4)
plot(-pi:0.001:pi,m)


%BRFT1LPFT1HPF.m
%BRF: Designed as the parallel realization of Type 1
%LPF (h(n)=h(N-1-n), N->Odd) and
% HPF (obtained using Type 1 LPF) with
%stop band between pi/8 and 3*pi/8
N=11;
wc1=pi/8;
wc2=3*pi/8;
tau=(N-1)/2;
n=0:1:N-1;
h1=sin(wc1*(n-tau))./(pi*(n-tau));
u=isnan(h1);
[p,q]=find(u==1);
h1(q)=wc1/pi;
n=0:1:N-1;
h2=sin(wc2*(n-tau))./(pi*(n-tau));
u=isnan(h2);
[p,q]=find(u==1);
h2(q)=wc2/pi;
h2=((-1*ones(1,length(h2))).^n).*h2;
h=h1+h2;
m=[];
for w=-pi:0.001:pi;
    m=[m abs(sum(h.*exp(-j*w*n)))];
```

```
end
figure
subplot(2,2,1)
plot(-pi:0.001:pi,m)


%BRFT3LPFT3HPF.m
%BRF: Designed as the parallel realization of Type 1
%LPF (h(n)=h(N-1-n), N->Even) and
% HPF (obtained using Type 3 LPF h(n)=h(N-1-n), N->Even)
% with stop band between pi/8 and 3*pi/8
N=10;
wc1=pi/8;
wc2=3*pi/8;
tau=(N-1)/2;
n=0:1:N-1;
h1=sin(wc1*(n-tau))./(pi*(n-tau));
u=isnan(h1);
[p,q]=find(u==1);
h1(q)=wc1/pi;
n=0:1:N-1;
h2=sin(wc2*(n-tau))./(pi*(n-tau));
u=isnan(h2);
[p,q]=find(u==1);
h2(q)=wc2/pi;
h2=((-1*ones(1,length(h2))).^n).*h2;
m=[];
n=0:1:length(h)-1;
for w=-pi:0.001:pi;
    m=[m abs(sum(h.*exp(-j*w*n)))];
end
subplot(2,2,2)
plot(-pi:0.001:pi,m)


%BRFT1LPFT3HPF.m
%BRF: Designed as the parallel realization of Type 1 LPF
%(h(n)=h(N-1-n), N->Odd) and HPF (obtained using Type 3 LPF)
%with stop band between pi/8 and 3*pi/8
N=11;
wc1=pi/8;
wc2=3*pi/8;
tau=(N-1)/2;
n=0:1:N-1;
h1=sin(wc1*(n-tau))./(pi*(n-tau));
u=isnan(h1);
[p,q]=find(u==1);
h1(q)=wc1/pi;
N=10;
n=0:1:N-1;
h2=sin(wc2*(n-tau))./(pi*(n-tau));
u=isnan(h2);
[p,q]=find(u==1);
h2(q)=wc2/pi;
h2=((-1*ones(1,length(h2))).^n).*h2;
h=h1+[h2 0];
m=[];
n=0:1:length(h)-1;
for w=-pi:0.001:pi;
    m=[m abs(sum(h.*exp(-j*w*n)))];
```

```
end
subplot(2,2,3)
plot(-pi:0.001:pi,m)
```

```
%BRFT3LPFT1HPF.m
%BRF: Designed as the parallel realization of Type 3 LPF
%(h(n)=h(N-1-n), N->Even) and HPF (obtained using Type 1 LPF
%h(n)=h(N-1-n), N->Odd) with stop band between pi/8 and 3*pi/8
N=10;
wc1=pi/8;
wc2=3*pi/8;
tau=(N-1)/2;
n=0:1:N-1;
h1=sin(wc1*(n-tau))./(pi*(n-tau));
u=isnan(h1);
[p,q]=find(u==1);
h1(q)=wc1/pi;
N=11;
n=0:1:N-1;
h2=sin(wc2*(n-tau))./(pi*(n-tau));
u=isnan(h2);
[p,q]=find(u==1);
h2(q)=wc2/pi;
h2=((-1*ones(1,length(h2))).^n).*h2;
h=[h1 0]+h2;
m=[];
n=0:1:length(h)-1;
for w=-pi:0.001:pi;
    m=[m abs(sum(h.*exp(-j*w*n)))];
end
subplot(2,2,4)
plot(-pi:0.001:pi,m)
```

3.2.4 Windows Used to Circumvent Ripples in the Magnitude Response

Truncated FIR filter is viewed as the product of the impulse response of the ideal FIR filter with the rectangular window. We also have observed that the ripples are introduced in the magnitude response of the truncated FIR filter (LPF, HPF, BPF and BRF). This is due to the convolution of ideal magnitude response and the magnitude response of the rectangular window (refer Fig. 3.10 for LPF). This is circumvented using other types of windows $w(n)$ (for the causal FIR filter obtained by shifting the noncausal impulse response toward right) as listed below (refer Fig. 3.19). The magnitude response (with N = 101) of the Low-pass filters, High-pass filter and Band pass filters and the Band reject filters are shown in the Figs. 3.21, 3.22, 3.23 and 3.24 respectively. It is seen from the figure that ripples in the magnitude response gets reduced if windowing (other than rectangular window) is used. Figure 3.20 shows the magnitude response for various windows with $N = 101$.

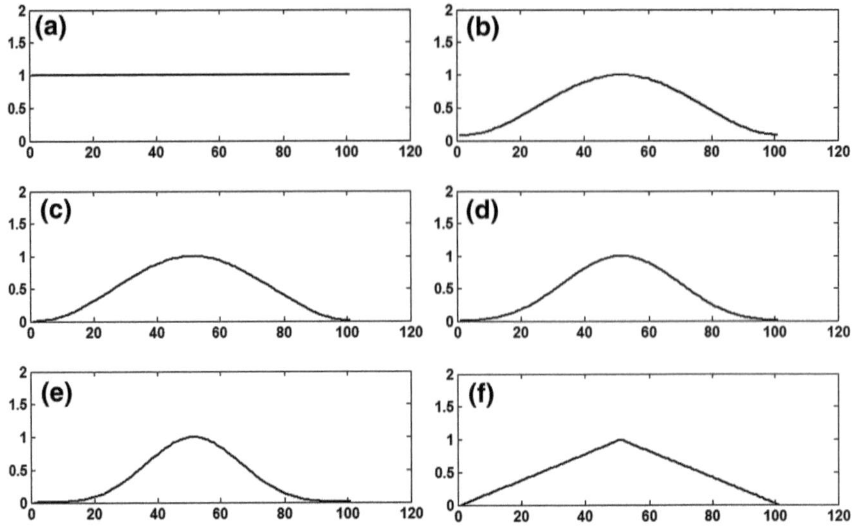

Fig. 3.19 Window functional values **a** Rectangular window. **b** Hamming window. **c** Hanning window. **d** Blackmannwindow1. **e** Blackmannwindow2. **f** Barlettwindow

- Rectangular window: $w(n) = 1$ for $0 \leq n \leq N-1$
- Hamming window: $w(n) = 0.54 - 0.46cos(\frac{2\pi n}{N})$ for $0 \leq n \leq N-1$
- Hanning window: $w(n) = 0.5 - 0.5cos(\frac{2\pi n}{N})$ for $0 \leq n \leq N-1$
- Blackmann window 1: $h(n) = 0.423 - 0.498cos(\frac{2\pi n}{N}) + 0.079cos(\frac{4\pi n}{N})$ for $0 \leq n \leq N-1$
- Blackmann window 2: $h(n) = 0.359 - 0.488cos(\frac{2\pi n}{N}) + 0.141cos(\frac{4\pi n}{N}) - 0.012 cos(\frac{6\pi n}{N})$ for $0 \leq n \leq N-1$
- Barlett window: $w(n) = 1 - \frac{2|n-\frac{N-1}{2}|}{(N-1)}$, for $0 \leq n \leq N-1$

```
%reduceripplesLPF.m
%Illustration of usage of window in reducing the ripples in Low-pass filter
N=101;
wc=pi/4;
tau=(N-1)/2;
n=0:1:N-1;
h=sin(wc*(n-tau))./(pi*(n-tau));
u=isnan(h);
[p,q]=find(u==1);
h(q)=wc/pi;
m=[];
for w=-pi:0.001:pi;
    m=[m abs(sum(h.*exp(-j*w*n)))];
end
%Windowing
rectwindow=ones(1,N);
hammingwindow= 0.54-0.46*cos(2*pi*n/N);
hanningwindow= 0.5-0.5*cos(2*pi*n/N);
blackmannwindow1= 0.423-0.498*cos(2*pi*n/N)+...
0.079*cos(4*pi*n/N);
```

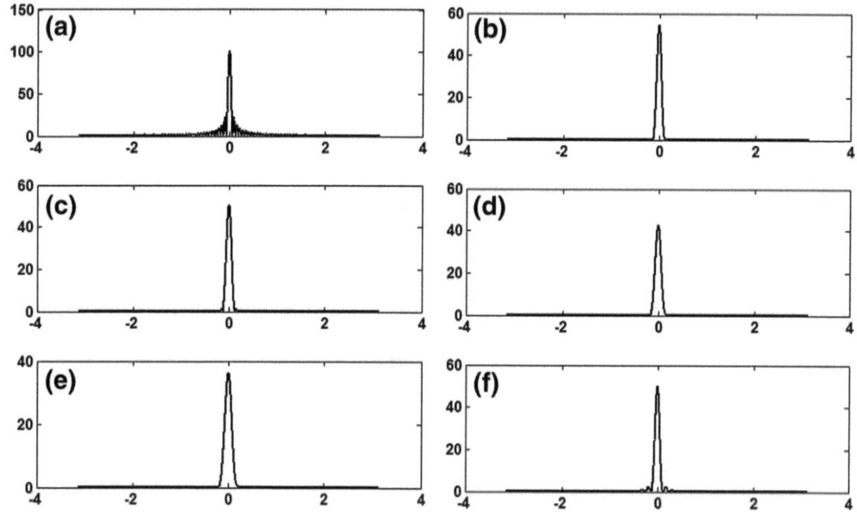

Fig. 3.20 Magnitude response of the Window function corresponding to Fig. 3.19 **a** Rectangular window. **b** Hamming window. **c** Hanning window. **d** Blackmannwindow1. **e** Blackmannwindow2. **f** Barlettwindow

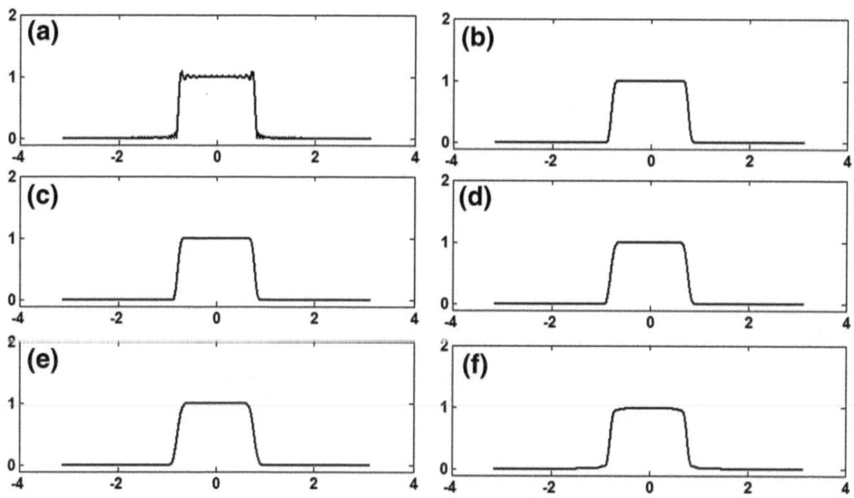

Fig. 3.21 Magnitude response of the corresponding (refer Fig. 3.20) low-pass filters (with cutoff frequency $\frac{\pi}{4}$) using **a** Rectangular window, **b** Hamming window, **c** Hanning window, **d** Blackmannwindow1, **e** Blackmannwindow2, **f** Barlettwindow

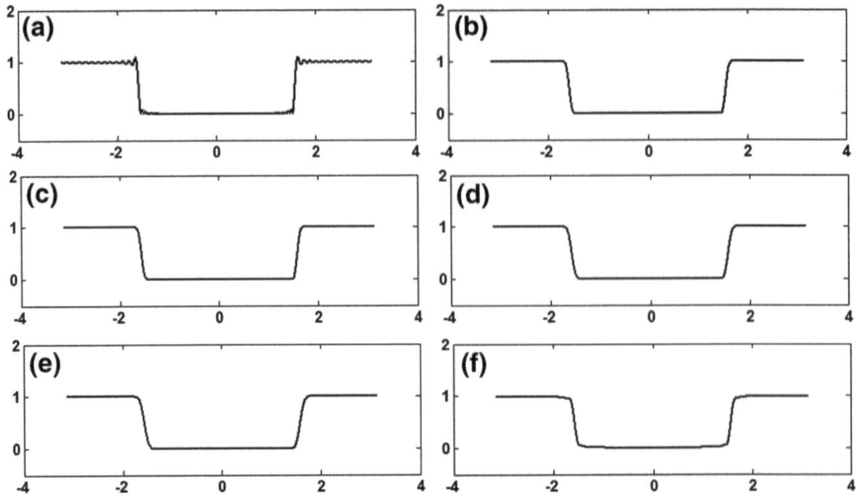

Fig. 3.22 Magnitude response of the corresponding (refer Fig. 3.20) high-pass filter (with cutoff frequency $\frac{\pi}{2}$) using **a** Rectangular window, **b** Hamming window, **c** Hanning window, **d** Blackmannwindow1, **e** Blackmannwindow2, **f** Barlettwindow

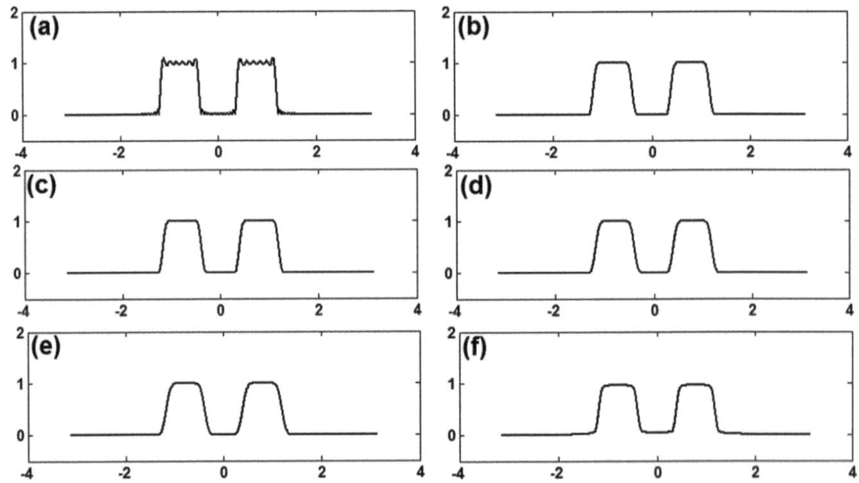

Fig. 3.23 Magnitude response of the corresponding (refer Fig. 3.20) band pass filter (with pass band range from $\frac{\pi}{8}$ to $\frac{3\pi}{8}$) using **a** Rectangular window, **b** Hamming window, **c** Hanning window, **d** Blackmannwindow1, **e** Blackmannwindow2, **f** Barlettwindow

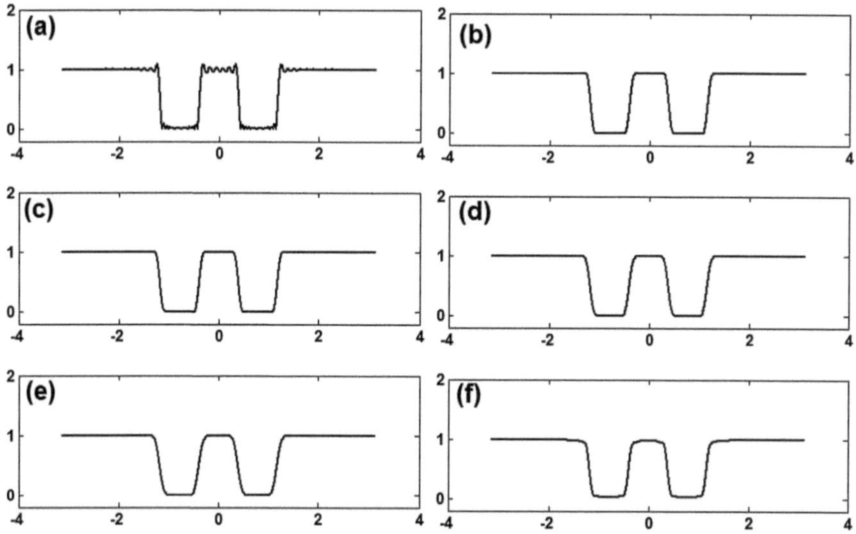

Fig. 3.24 Magnitude response of the corresponding (refer Fig. 3.20) band reject filter (with stop band range from $\frac{\pi}{8}$ to $\frac{3\pi}{8}$) using **a** Rectangular window, **b** Hamming window, **c** Hanning window, **d** Blackmannwindow1, **e** Blackmannwindow2, **f** Barlettwindow

```
blackmannwindow2= 0.359-0.488*cos(2*pi*n/N)+...
0.141*cos(4*pi*n/N)-0.012*cos(6*pi*n/N);
barlettwindow=1-2*abs((n-(N-1)/2))/(N-1);
%Magnitude response of various windows
Mrect=[];
Mhamm=[];
Mhann=[];
Mblack1=[];
Mblack2=[];
Mbarlett=[];
for w=-pi:0.001:pi;
    Mrect=[Mrect abs(sum(rectwindow.*exp(-j*w*n)))];
    Mhamm=[Mhamm abs(sum(hammingwindow.*exp(-j*w*n)))];
    Mhann=[Mhann abs(sum(hanningwindow.*exp(-j*w*n)))];
    Mblack1=[Mblack1 abs(sum(blackmannwindow1.*exp(-j*w*n)))];
    Mblack2=[Mblack2 abs(sum(blackmannwindow2.*exp(-j*w*n)))];
    Mbarlett=[Mbarlett abs(sum(barlettwindow.*exp(-j*w*n)))];
end
figure
subplot(3,2,1)
plot(rectwindow);
subplot(3,2,2)
plot(hammingwindow);
subplot(3,2,3)
plot(hanningwindow);
subplot(3,2,4)
plot(blackmannwindow1);
subplot(3,2,5)
plot(blackmannwindow2);
subplot(3,2,6)
plot(barlettwindow);
```

```
figure
subplot(3,2,1)
plot(-pi:0.001:pi,Mrect);
subplot(3,2,2)
plot(-pi:0.001:pi,Mhamm);
subplot(3,2,3)
plot(-pi:0.001:pi,Mhann);
subplot(3,2,4)
plot(-pi:0.001:pi,Mblack1);
subplot(3,2,5)
plot(-pi:0.001:pi,Mblack2);
subplot(3,2,6)
plot(-pi:0.001:pi,Mbarlett);


hamm=h.*hammingwindow;
hann=h.*hammingwindow;
black1=h.*blackmannwindow1;
black2=h.*blackmannwindow2;
barlett=h.*barlettwindow;
mhamm=[];
mhann=[];
mblack1=[];
mblack2=[];
mbarlett=[];
for w=-pi:0.001:pi;
    mhamm=[mhamm abs(sum(hamm.*exp(-j*w*n)))];
    mhann=[mhann abs(sum(hann.*exp(-j*w*n)))];
    mblack1=[mblack1 abs(sum(black1.*exp(-j*w*n)))];
    mblack2=[mblack2 abs(sum(black2.*exp(-j*w*n)))];
    mbarlett=[mbarlett abs(sum(barlett.*exp(-j*w*n)))];
end
figure
subplot(3,2,1)
plot(-pi:0.001:pi,m);
subplot(3,2,2)
plot(-pi:0.001:pi,mhamm);
subplot(3,2,3)
plot(-pi:0.001:pi,mhann);
subplot(3,2,4)
plot(-pi:0.001:pi,mblack1);
subplot(3,2,5)
plot(-pi:0.001:pi,mblack2);
subplot(3,2,6)
plot(-pi:0.001:pi,mbarlett);



%reduceripplesLPF.m
%Illustration of usage of window in
%reducing the ripples in High-pass filter
N=101;
wc=pi/2;
tau=(N-1)/2;
n=0:1:N-1;
h=sin(wc*(n-tau))./(pi*(n-tau));
u=isnan(h);
[p,q]=find(u==1);
h(q)=wc/pi;
h=((-1*ones(1,length(h))).^n).*h;
```

```
hamm=h.*hammingwindow;
hann=h.*hammingwindow;
black1=h.*blackmannwindow1;
black2=h.*blackmannwindow2;
barlett=h.*barlettwindow;
m=[];
mhamm=[];
mhann=[];
mblack1=[];
mblack2=[];
mbarlett=[];
for w=-pi:0.001:pi;
    m=[m abs(sum(h.*exp(-j*w*n)))];
    mhamm=[mhamm abs(sum(hamm.*exp(-j*w*n)))];
    mhann=[mhann abs(sum(hann.*exp(-j*w*n)))];
    mblack1=[mblack1 abs(sum(black1.*exp(-j*w*n)))];
    mblack2=[mblack2 abs(sum(black2.*exp(-j*w*n)))];
    mbarlett=[mbarlett abs(sum(barlett.*exp(-j*w*n)))];
end
figure
subplot(3,2,1)
plot(-pi:0.001:pi,m);
subplot(3,2,2)
plot(-pi:0.001:pi,mhamm);
subplot(3,2,3)
plot(-pi:0.001:pi,mhann);
subplot(3,2,4)
plot(-pi:0.001:pi,mblack1);
subplot(3,2,5)
plot(-pi:0.001:pi,mblack2);
subplot(3,2,6)
plot(-pi:0.001:pi,mbarlett);
```

```
%reduceripplesBPF.m
%Illustration of usage of window
%in reducing the ripples in Band pass
%filter between pi/8 and 3pi/8
N=101;
wc1=pi-pi/8;
wc2=3*pi/8;
tau-(N-1)/2;
n=0:1:N-1;
h1=sin(wc2*(n-tau))./(pi*(n-tau));
u=isnan(h1);
[p,q]=find(u==1);
h1(q)=wc2/pi;
hammh1=h1.*hammingwindow;
hannh1=h1.*hammingwindow;
black1h1=h1.*blackmannwindow1;
black2h1=h1.*blackmannwindow2;
barletth1=h1.*barlettwindow;

N=101;
n=0:1:N-1;
h2=sin(wc1*(n-tau))./(pi*(n-tau));
u=isnan(h2);
[p,q]=find(u==1);
h2(q)=wc1/pi;
h2=((-1*ones(1,length(h2))).^n).*h2;
```

```
hammh2=h2.*hammingwindow;
hannh2=h2.*hammingwindow;
black1h2=h2.*blackmannwindow1;
black2h2=h2.*blackmannwindow2;
barletth2=h2.*barlettwindow;

h=conv(h1,h2);
hamm=conv(hammh1,hammh2);
hann=conv(hannh1,hannh2);
black1=conv(black1h1,black1h2);
black2=conv(black2h1,black2h2);
barlett=conv(barletth1,barletth2);
n=0:1:length(h)-1;
m=[];
mhamm=[];
mhann=[];
mblack1=[];
mblack2=[];
mbarlett=[];
for w=-pi:0.001:pi;
    m=[m abs(sum(h.*exp(-j*w*n)))];
    mhamm=[mhamm abs(sum(hamm.*exp(-j*w*n)))];
    mhann=[mhann abs(sum(hann.*exp(-j*w*n)))];
    mblack1=[mblack1 abs(sum(black1.*exp(-j*w*n)))];
    mblack2=[mblack2 abs(sum(black2.*exp(-j*w*n)))];
    mbarlett=[mbarlett abs(sum(barlett.*exp(-j*w*n)))];
end
figure
subplot(3,2,1)
plot(-pi:0.001:pi,m);
subplot(3,2,2)
plot(-pi:0.001:pi,mhamm);
subplot(3,2,3)
plot(-pi:0.001:pi,mhann);
subplot(3,2,4)
plot(-pi:0.001:pi,mblack1);
subplot(3,2,5)
plot(-pi:0.001:pi,mblack2);
subplot(3,2,6)
plot(-pi:0.001:pi,mbarlett);
```

```
%reduceripplesBRF.m
%Illustration of usage of window in
%reducing the ripples in Band reject
%filter between pi/8 and 3*pi/8
N=101;
wc1=pi/8;
wc2=pi-3*pi/8;
tau=(N-1)/2;
n=0:1:N-1;
h1=sin(wc1*(n-tau))./(pi*(n-tau));
u=isnan(h1);
[p,q]=find(u==1);
h1(q)=wc1/pi;
hammh1=h1.*hammingwindow;
hannh1=h1.*hammingwindow;
black1h1=h1.*blackmannwindow1;
black2h1=h1.*blackmannwindow2;
barletth1=h1.*barlettwindow;
```

```
N=101;
n=0:1:N-1;
h2=sin(wc2*(n-tau))./(pi*(n-tau));
u=isnan(h2);
[p,q]=find(u==1);
h2(q)=wc2/pi;
h2=((-1*ones(1,length(h2))).^n).*h2;
hammh2=h2.*hammingwindow;
hannh2=h2.*hammingwindow;
black1h2=h2.*blackmannwindow1;
black2h2=h2.*blackmannwindow2;
barletth2=h2.*barlettwindow;

h=h1+h2;
hamm=hammh1+hammh2;
hann=hannh1+hannh2;
black1=black1h1+black1h2;
black2=black2h1+black2h2;
barlett=barletth1+barletth2;

n=0:1:length(h)-1;
m=[];
mhamm=[];
mhann=[];
mblack1=[];
mblack2=[];
mbarlett=[];
for w=-pi:0.001:pi;
    m=[m abs(sum(h.*exp(-j*w*n)))];
    mhamm=[mhamm abs(sum(hamm.*exp(-j*w*n)))];
    mhann=[mhann abs(sum(hann.*exp(-j*w*n)))];
    mblack1=[mblack1 abs(sum(black1.*exp(-j*w*n)))];
    mblack2=[mblack2 abs(sum(black2.*exp(-j*w*n)))];
    mbarlett=[mbarlett abs(sum(barlett.*exp(-j*w*n)))];
end
figure
subplot(3,2,1)
plot(-pi:0.001:pi,m);
subplot(3,2,2)
plot(-pi:0.001:pi,mhamm);
subplot(3,2,3)
plot(-pi:0.001:pi,mhann);
subplot(3,2,4)
plot(-pi:0.001:pi,mblack1);
subplot(3,2,5)
plot(-pi:0.001:pi,mblack2);
subplot(3,2,6)
plot(-pi:0.001:pi,mbarlett);
```

3.3 FIR Filters that Have Identical Magnitude Response

Let the zeros of the FIR filter ($H_1(z)$) with real coefficients are represented as z_1, z_1^*, z_2 and z_2^*. Note that the roots are available as conjugate the pair. The other three FIR filter with real coefficients that has identical magnitude response as that of the $H_1(z)$

are listed below.

$$Roots\ of\ H_2(z) : z_1, z_1^*, \frac{1}{z_2} and \frac{1}{z_2^*}$$
$$Roots\ of\ H_3(z) : z_1, z_1^*, \frac{1}{z_2} and \frac{1}{z_2^*}$$
$$Roots\ of\ H_4(z) : \frac{1}{z_1}, \frac{1}{z_1^*}, \frac{1}{z_2} and \frac{1}{z_2^*}$$

The typical example of the Transfer function of four FIR filters (with $z_1 = 1.5 + 1.5j, z_2 = 2.5 + 2.5j$) that have the identical magnitude response and different phase responses are listed below.

$$H_1(z) = 1 - 8z^{-1}32z^{-2} - 60z^{-3}56.2500z^{-4}$$
$$H_2(z) = 1 - 3.4z^{-1} + 5.78z^{-2} - 2.04z^{-3}56.2500z^{-4}$$
$$H_3(z) = 1 - 5.67z^{-1} + 16.05z^{-2} - 9.44z^{-3} + 2.778z^{-4}$$
$$H_4(z) = 1 - 8z^{-1}32z^{-2} - 60z^{-3}56.2500z^{-4}$$

The corresponding pole-zero plot are shown in the Fig. 3.25. From Figs. 3.26 and 3.27, it is observed that the magnitude response of the four FIR filters are identical and the corresponding phase responses are distinct.

It is also observed that the magnitude response of the FIR filter with zero r_z is having identical magnitude response with that of the FIR filter with zero $\frac{1}{r_z^*}$ (refer Fig. 3.28).

```
%identicalmagresfilters.m
r1=1.5+1.5j; r2=2.5+2.5j;
h1=[r1 r1' r2 r2'];POLY1=poly(h1);
figure(1)
subplot(2,2,1)
zplane(POLY1,1)
figure(2)
subplot(2,2,1)
plot(abs(fft(POLY1,1000)))
figure(3)
subplot(2,2,1)
plot(angle(fft(POLY1,1000)))

h2=[r1 r1' 1/r2 1/r2'];
POLY2=poly(h2);
roots(POLY2)
figure(1)
subplot(2,2,2)
zplane(POLY2,1)
figure(2)
subplot(2,2,2)
plot(abs(fft(POLY2,1000)))
figure(3)
```

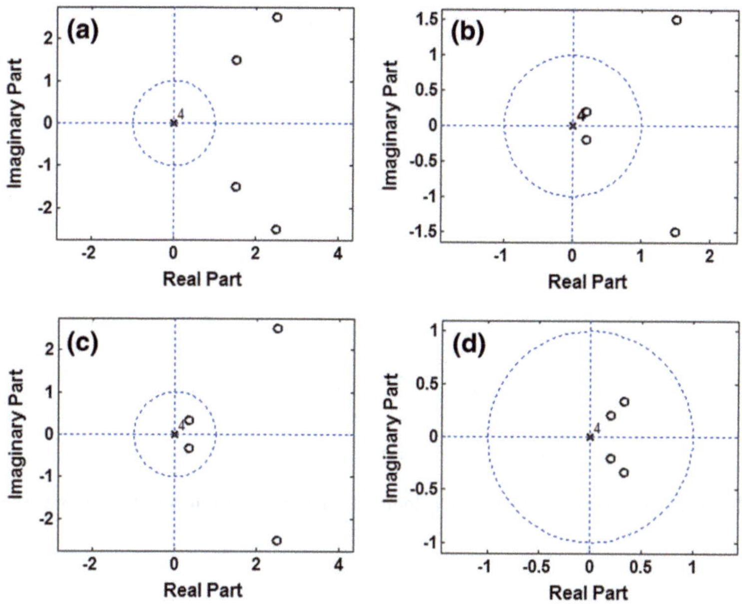

Fig. 3.25 Pole-zero plot of four FIR filters with real coefficient that have identical magnitude response

```
subplot(2,2,2)
plot(angle(fft(POLY2,1000)))

h3=[1/r1 1/r1' r2 r2'];
POLY3=poly(h3);
figure(1)
subplot(2,2,3)
zplane(POLY3,1)
figure(2)
subplot(2,2,3)
plot(abs(fft(POLY3,1000)))
figure(3)
subplot(2,2,3)
plot(angle(fft(POLY3,1000)))

h4=[1/r1 1/r1' 1/r2 1/r2'];
POLY4=poly(h4);
figure(1)
subplot(2,2,4)
zplane(POLY4,1)
figure(2)
subplot(2,2,4)
plot(abs(fft(POLY4,1000)))
figure(3)
subplot(2,2,4)
plot(angle(fft(POLY4,1000)))
```

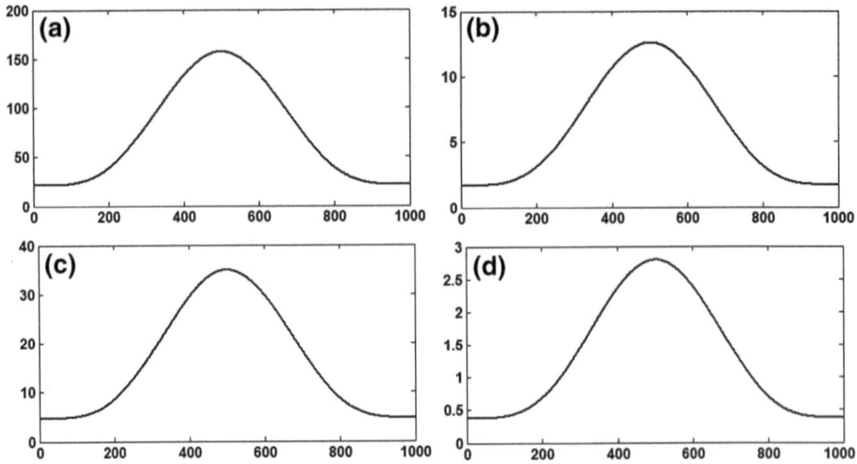

Fig. 3.26 The corresponding magnitude response of the four FIR filters (refer Fig. 3.25)

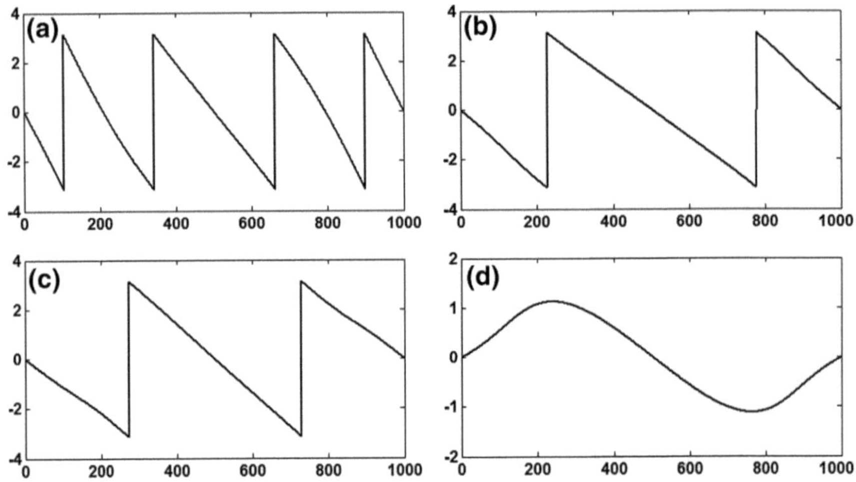

Fig. 3.27 The corresponding phase response of the four FIR filters (refer Fig. 3.25)

It is seen from the Fig. 3.28d ($H_4(z)$), that all the zeros lie within the unit circle. This corresponds to minimum phase filter. Also from Fig. 3.28a ($H_1(z)$), that all the zeros lie outside the unit circle. This corresponds to maximum phase filter. The transfer function $H_1(z)$ (min-phase) and $H_2(z)$ (max-phase) are related using the all pass filter $H_{allpass}(z)$ as described below.

$$H_2(z) = H_1(z)H_{allpass}(z) \tag{3.16}$$

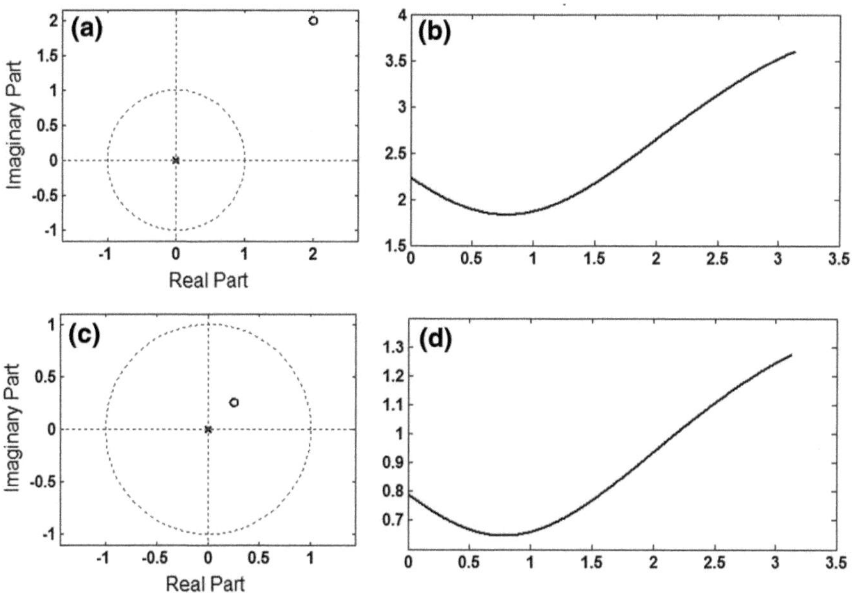

Fig. 3.28 **a** Pole-zero plot with zero at $2 + 2j$, **b** Corresponding magnitude response, **c** Pole-zero plot with zero at $\frac{1}{2+2j}$, **d** Corresponding magnitude response

The pole-zero plot of the All pass filter, Minimum phase filter and the Maximum phase filter are given in the Fig. 3.29 and the corresponding magnitude response is given in Fig. 3.30.

```
r1=1.5+1.5j; r2=2.5+2.5j;
h=[r1 r1' r2 r2'];
APFzeros=[1/r1 1/r1' 1/r2 1/r2'];
APFpoles=[r1 r1' r2 r2'];
POLY1NUM=poly(APFpoles);
POLY1DEN=poly(APFzeros);

figure(1)
subplot(1,3,2)
zplane(POLY1NUM,POLY1DEN)

figure(2)
subplot(1,3,2)
[H,W]=freqz(POLY1NUM,POLY1DEN);
plot(W,abs(H))
%%%%%%%%%%%%%%%%%%%%%%%%%%%%%%%
h4=[1/r1 1/r1' 1/r2 1/r2'];
POLY4=poly(h4);
figure(1)
subplot(1,3,3)
zplane(POLY4,1)

figure(2)
```

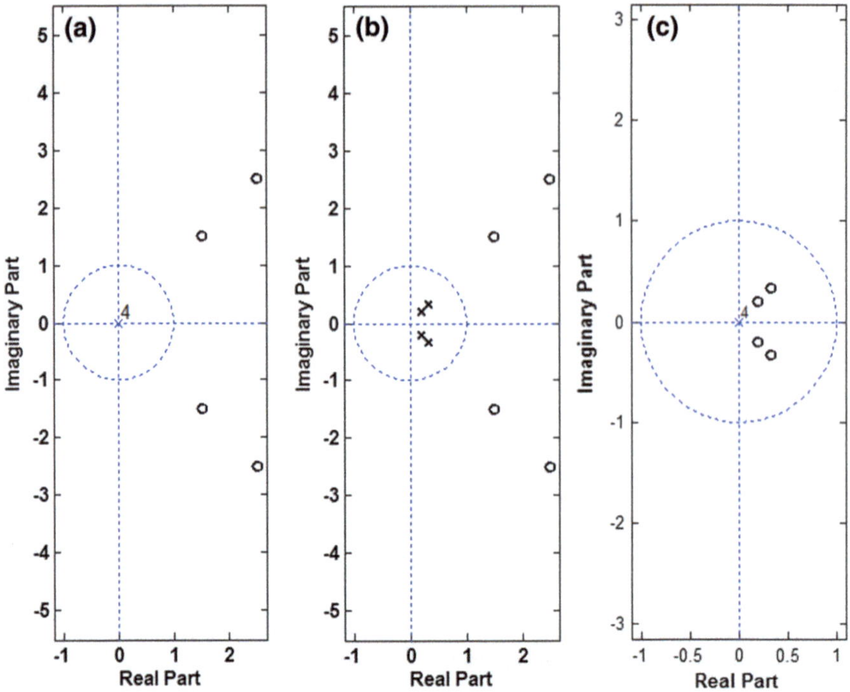

Fig. 3.29 Pole zero plot: **a** maximum phase filter, **b** all pass filter, **c** minimum phase filter

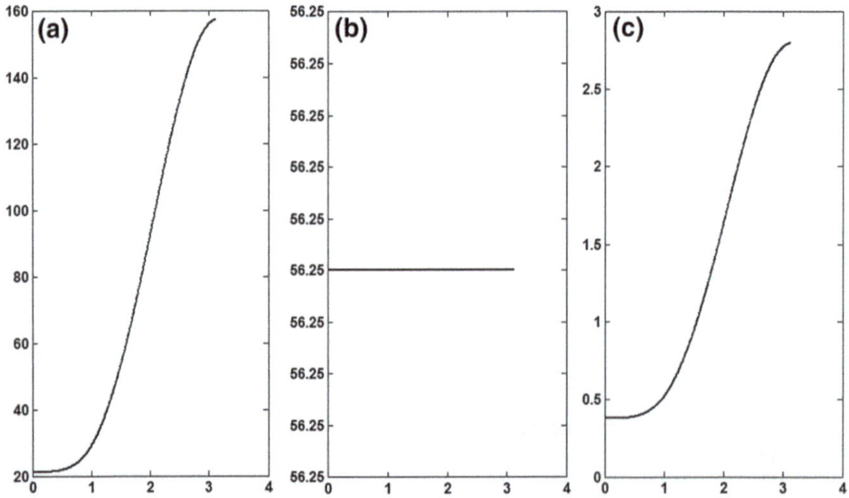

Fig. 3.30 Magnitude response corresponding to Fig. 3.29 **a** maximum phase filter, **b** all pass filter, **c** minimum phase filter

```
subplot(1,3,3)
[H,W]=freqz(POLY4,1);
plot(W,abs(H))
%%%%%%%%%%%%%%%%%%%%%%%%%%%
h1=[r1 r1' r2 r2'];
POLY1=poly(h1);
figure(1)
subplot(1,3,1)
zplane(POLY1,1)

figure(2)
subplot(1,3,1)
[H,W]=freqz(POLY1,1);
plot(W,abs(H))
%%%%%%%%%%%%%%%%%%%%%%%%%%
```

Chapter 4
Multirate Digital Signal Processing

4.1 Sampling Rate Conversion by the Factor $\frac{M}{N}$

Let the band limited signal $x(t)$ is sampled with the sampling frequency Fs to obtain the sequence $x(n)$. The method of getting the new sequence yn obtained by sampling the signal $x(t)$ with sampling frequency $Fs_{new} = \frac{M}{N}F_s$ is called sampling rate conversion. This is done by the following.

4.1.1 Upsampling (with $N = 1$)

Let the sequence $x(n)$ be represented as $[x(0)\ x(1)\ x(2)\ x(3)\ x(4)\ x(5)\ x(6)\cdots]$. The output sequence after up sampling is given by
$y = [x(0)\ 0\ x(1)\ 0\ x(2)\ 0\ x(3)\ 0\ x(4)\ 0\ x(5)\ 0\ x(6)\cdots]$. Let the z-transformation of the sequence $x(n)$ be represented as $X(z)$. The z-transformation of the sequence $y(n)$ is obtained as follows:

$$X(z) = x(0) + x(1)z^{-1} + x(2)z^{-2} + x(3)z^{-3} + x(4)z^{-4} + x(5)z^{-5} + x(6)z^{-6}\cdots$$
$$Y(z) = y(0) + y(1)z^{-2} + y(2)z^{-2} + y(3)z^{-3} + y(4)z^{-4} + y(5)z^{-5}\cdots$$
$$\Rightarrow Y(z) = x(0) + 0z^{-1} + x(1)z^{-2} + 0z^{-3} + x(2)z^{-4} + 0z^{-5} + x(3)z^{-6} + \cdots$$
$$\Rightarrow Y(z) = X(z^2).$$

In general, the z-transformation of the $Y(z)$ with arbitrary M with $N = 1$ is given as $Y(z) = X(z^M)$.

E.S. Gopi, *Multi-Disciplinary Digital Signal Processing*,
DOI 10.1007/978-3-319-57430-1_4

4.1.2 Downsampling (with $M = 1$)

Let the sequence $x(n)$ be represented as $[x(0)\ x(1)\ x(2)\ x(3)\ x(4)\ x(5)\ x(6)\cdots]$. The output sequence after down sampling is given by $y(n) = [x(0)\ x(2)\ x(4)\ x(6)\ ]$. Let the z-transformation of $x(n)$ is $X(z)$ and the z-transformation of the sequence $y(n)$ is computed as follows.

$$Y(z) = y(0) + y(1)z^{-1} + y(2)z^{-2} + y(3)z^{-3} + y(4)z^{-4} + y(5)z^{-5} + y(6)z^{-6} + y(7)z^{-7} + \cdots$$
$$\Rightarrow Y(z) = x(0) + x(2)z^{-1} + x(4)z^{-2} + x(6)z^{-3} + x(8)z^{-4} + x(10)z^{-5} + x(12)z^{-6} + x(14)z^{-6} \ldots$$

Consider the new sequence $w(n) = x(n)z(n)$, where $z(n) = [1\cdots0\cdots1\cdots 0\cdots]$ and hence $w(n)$ is obtained as follows:

$$\begin{aligned} W(z) &= w(0) + w(1)z^{-1} + w(2) + z^{-2} + \\ \Rightarrow W(z) &= x(0) + 0z^{-1} + x(2)z^{-2} + \cdots \\ W(z^{1/2}) &= x(0) + x(2)z^{-1} + x(4)z^{-2} + \cdots \\ &= Y(z). \end{aligned}$$

Thus if $W(z)$ is computed, then $Y(z)$ is obtained as $Y(z) = W(z^{1/2})$. Using IDFT, we represent $w(n)$ as follows.

$$w(n) = x(n)\frac{1}{2}\sum_{k=0}^{k=1} e^{\frac{j2\pi kn}{2}} \tag{4.1}$$

The z-transformation is obtained as follows:

$$W(z) = \sum_{n=0}^{n=\infty} w(n)z^{-n} \tag{4.2}$$

$$= \sum_{n=0}^{n=\infty} x(n)\frac{1}{2}\sum_{k=0}^{k=1} e^{\frac{j2\pi kn}{2}} z^{-n} \tag{4.3}$$

$$= \frac{1}{2}\sum_{k=0}^{k=1}\sum_{n=0}^{n=\infty} x(n)e^{\frac{j2\pi kn}{2}} z^{-n} \tag{4.4}$$

$$= \frac{1}{2}\sum_{k=0}^{k=1}\sum_{n=0}^{n=\infty} x(n)(e^{-\frac{j2\pi k}{2}} z)^{-n} \tag{4.5}$$

$$= \frac{1}{2}\sum_{k=0}^{k=1} X(e^{-\frac{j2\pi k}{2}} z) \tag{4.6}$$

$$= \frac{1}{2}(X(z) + X(-z)). \tag{4.7}$$

Thus $Y(z)$ is computed as follows:

$$Y(z) = W(z^{1/2}) = \frac{1}{2}(X(z^{1/2}) + X(-z^{1/2})). \tag{4.8}$$

In general for arbitrary value of N, with $M = 1$, we get the following:

$$Y(z) = \frac{1}{N}\sum_{k=0}^{k=N-1} X(e^{\frac{j2\pi k}{N}} z^{\frac{1}{N}}) \tag{4.9}$$

$$= \frac{1}{N}\sum_{k=0}^{k=N-1} X(e^{\frac{j2\pi k}{N}} e^{\frac{jw}{N}}) \tag{4.10}$$

$$= \frac{1}{N}\sum_{k=0}^{k=N-1} X(e^{\frac{j(2\pi k+w)}{N}}). \tag{4.11}$$

4.1.3 Comments on the Fig. 4.1

1. The graph is plotted between frequency in radians in the x-axis and the corresponding magnitude in the y-axis.
2. Subplot 1: In the first subplot, 2π in the x-axis corresponds to $F_s = \frac{F_{max}}{2}$ Hz, where F_{max} is the allowed maximum frequency content (satisfying the sampling theorem) of the signal. F_1 is the actual maximum frequency content of the signal
3. Subplot 2: In the second subplot, 2π in the x-axis corresponds to $\frac{F_s}{2} = \frac{F_{max}}{4}$ Hz, where F_{max} is the allowed maximum frequency content (satisfying the sampling theorem) of the signal. F_1 is the actual maximum frequency content of the signal. Note that the maximum frequency allowed in this case has been reduced from $\frac{F_{max}}{2}$ to $\frac{F_{max}}{4}$. Suppose if the actual F_1 is greater than $\frac{F_{max}}{2}$, overlapping occur. This is circumvented by passing the signal through the Low-pass filter with cutoff frequency $\frac{F_{max}}{2} = \frac{\pi}{2}$(corresponding to the first subplot), before subjected to down sampling.
4. Subplot 3: In the third subplot, 2π in the x-axis corresponds to $2F_s = F_{max}$ Hz, where F_{max} is the allowed maximum frequency content (satisfying the sampling theorem) of the signal. F_1 is the actual maximum frequency content of the signal. Note that the maximum frequency allowed in this case has been increased from $\frac{F_{max}}{2}$ to F_{max}. Hence there is no need to filter the signal before performing down sampling. But there exist the mirror image of the original spectrum (shaded region). Hence the spectrum looks like the with cutoff frequency $\frac{\pi}{2}$(corresponding to the third subplot).

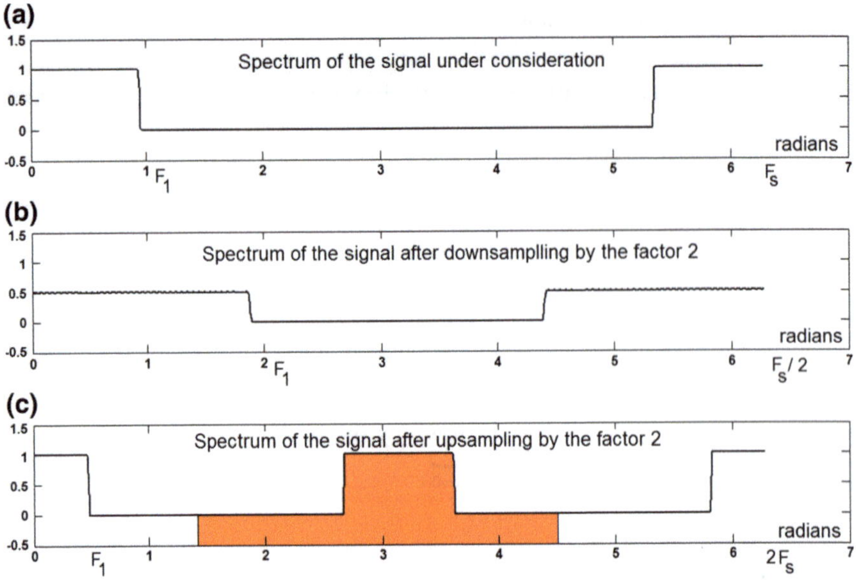

Fig. 4.1 Illustration of decimation and interpolation. **a** Spectrum of the signal under consideration. **b** Spectrum of the signal after down sampling by the factor 2. **c** Spectrum of the corresponding signal after up sampling by the factor 2

```
%multiratedemo.m
%Generating the input sequence
x=fir1(1000,0.3);
MAG=[];
n=0:1:1000;
for w=0:0.01:2*pi
s=sum(x.*exp(-j*w*n));
MAG=[MAG s];
end
%Downsampling by 2
y=dyaddown(x,2);
n=0:1:length(y)-1;
MAGD=[];
for w=0:0.01:2*pi
s=sum(y.*exp(-j*w*n));
MAGD=[MAGD s];
end
%upsampling by 2
z=dyadup(x,2);
n=0:1:length(z)-1;
MAGU=[];
for w=0:0.01:2*pi
s=sum(z.*exp(-j*w*n));
MAGU=[MAGU s];
end
```

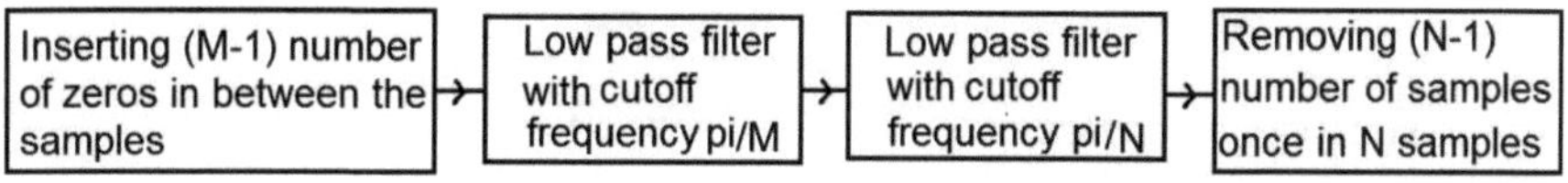

Fig. 4.2 Block diagram illustrating the steps involved in sampling rate conversion by the factor $\frac{M}{N}$

```
figure
subplot(3,1,1)
plot(linspace(0,2*pi,length(MAG)),abs(MAG),'-')
subplot(3,1,2)
plot(linspace(0,2*pi,length(MAG)),abs(MAGD),'-')
subplot(3,1,3)
plot(linspace(0,2*pi,length(MAG)),abs(MAGU),'-')
```

In general, the steps involved in getting the sequence y_n (that is obtained by sampling the signal $x(t)$ with the sampling frequency $Fs_{new} = \frac{M}{N}F_s$) from the sequence x_n (that is obtained by sampling the signal $x(t)$ with the sampling frequency $Fs_{new} = \frac{M}{N}F_s$) are summarized below (refer Fig. 4.2).

1. Up sample the sequence by inserting $(M-1)$ zeros in between every samples.
2. The obtained sequence is subjected to the high-pass filtering with cutoff frequency $\frac{\pi}{M}$ to get the interpolated sequence.
3. The up sampled sequence is further passed through low-pass filter with cutoff frequency $\frac{\pi}{N}$.
4. The sequence obtained is further down sampled by removing $(N-1)$ samples once in N samples.
5. The two low-pass filters used in this sampling rate conversion can be combined as the single Low-pass filter with cutoff frequency chosen as the minimum of $\frac{\pi}{M}$ and $\frac{\pi}{N}$.

4.2 Poly-phase Realization of the Filter for Interpolation

Interpolation by M is obtained by up sampling the sequence by M, i.e., inserting $(M-1)$ zeros between the samples to get the new sequence. It is followed by passing the sequence through the Low-pass FIR filter with cutoff $\frac{pi}{M}$. This is realized by constructing the M poly-phase filters as described below (with M = 3).

1. Consider the order of the filter (say 6) as the integer multiples of $M = 3$.
2. The impulse response of the filter is given as $h = [h_0\ h_1\ h_2\ h_3\ h_4\ h_5$. Note that the cutoff frequency is $\frac{pi}{M} = \frac{pi}{3}$. Obtain the three (in general M) polyfilters as $poly1 = [h_0\ h_3]$, $poly2 = [h_1\ h_4]$, and $poly3 = [h_2\ h_5]$.
3. It is observed that the $poly1$, $poly2$, and $poly3$ are the downsampled version of the filter h.

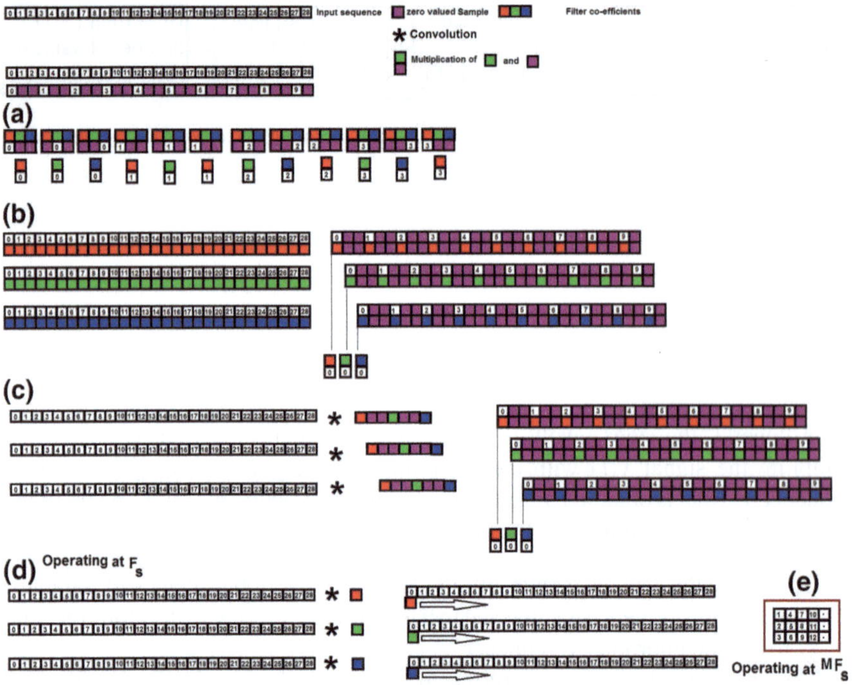

Fig. 4.3 Illustration of interpolation by 3 using the Interpolator filter (order 3). **a** Direct method. **b** Efficient method. **c** Practical implementation of the efficient method. **d** Poly-phase method. **e** The order in which the output is obtained in the poly-phase method

4. We have already identified that if the filter coefficients are downsampled, we get the average of shifted and scaled up version of the M original magnitude response (refer (4.11)). In particular if the cutoff frequency of the filter is $\frac{\pi}{M}$, we get the magnitude response as the flat (All pass).
5. Thus $poly1$, $poly2$, and $poly3$ are having flat magnitude response (All pass filter) (refer Fig. 4.5b–d)with different phase response. Hence the filter is named as poly-phase filter. The actual magnitude response of the filter is given in Fig. 4.5a.
6. Filtering the sequence (obtained after up sampling) through the FIR filter is identical with filtering through three poly-phase filters and collecting the outputs in the cyclic manner (refer Figs. 4.3 and 4.4). The order in which the output is collected is illustrated in Figs. 4.3e and 4.4e.
7. The spectrum of the sequence before and after interpolation using the direct filtering method and using poly-phase technique Fig. 4.5 is illustrated in Fig. 4.6.

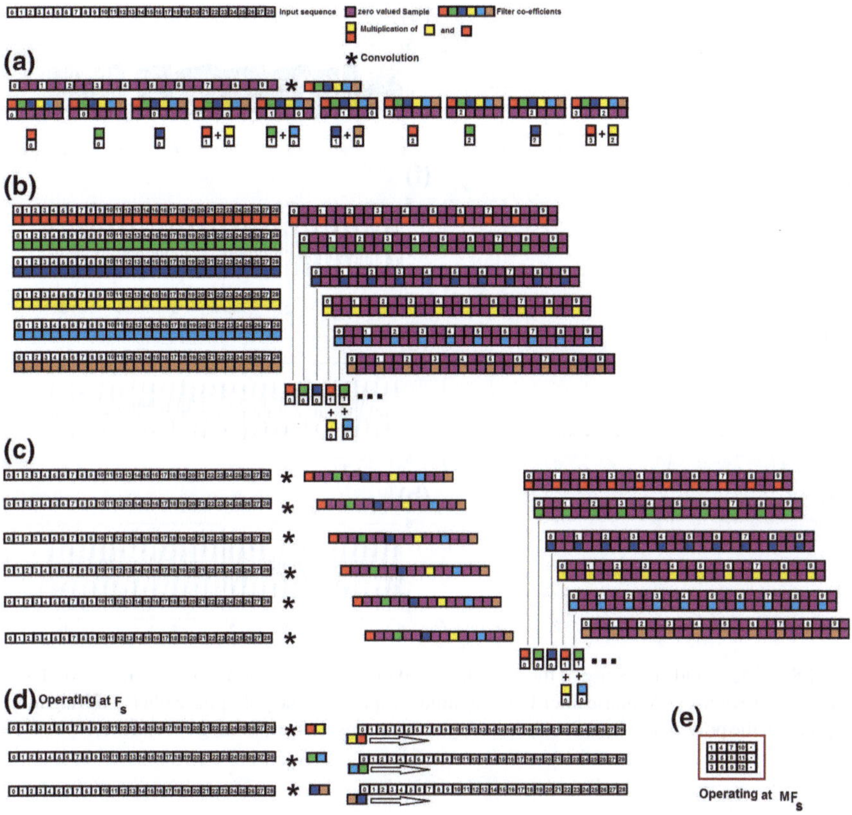

Fig. 4.4 Illustration of interpolation by 3 using the Interpolator filter (order 6). **a** Direct method. **b** Efficient method. **c** Practical implementation of the efficient method. **d** Poly-phase method. **e** The order in which the output is obtained in the poly-phase method

```
%interpolyphase.m
%Polyphase realization of sampling rate conversion
%(Interpolation) by the factor 3
%Generating the input sequence
x=fir1(1000,1/6); %Not the filter coefficients.
MAG=[];
n=0:1:1000;
for w=0:0.01:2*pi
s=sum(x.*exp(-j*w*n));
MAG=[MAG s];
end

%Interpolation by the factor 3
z=[x;zeros(2,length(x))]
z=reshape(z,1,size(z,1)*size(z,2));
```

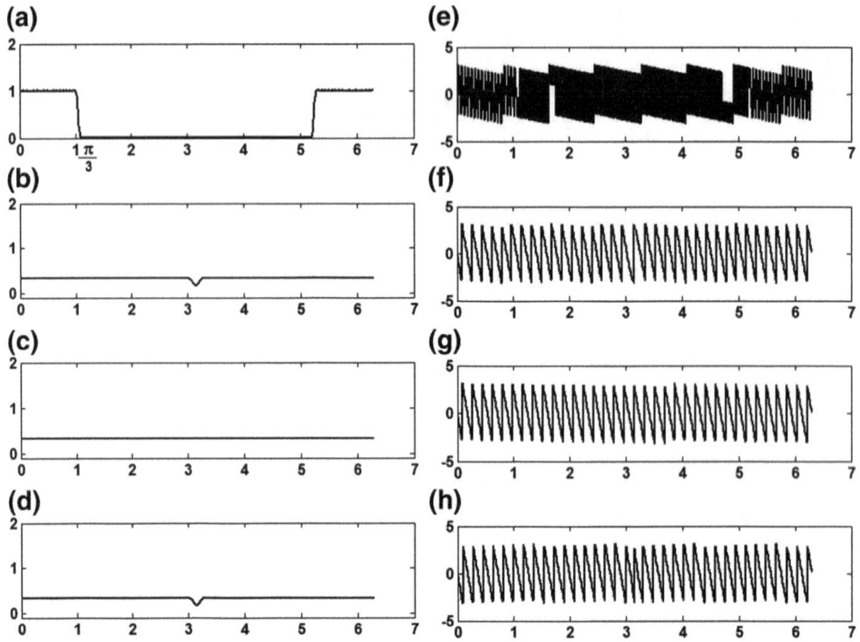

Fig. 4.5 **a** Magnitude response of the FIR filter used in the interpolation (direct method). **b** Magnitude response of the poly-phase filter 1. **c** Magnitude response of the poly-phase filter 2. **d** Magnitude response of the poly-phase filter 3. **e**–**h** Are the phase response corresponding to (**a**)–(**d**), respectively

```
%We need to filter the obtained sequence
%using low-pass filter with cutoff
%frequency pi/3;
h=fir1(212,1/3);
y=conv(z,h)
%Realization using poly-phase filter.
hdyad=reshape(h,3,length(h)/3);
h1=hdyad(1,:);
h2=hdyad(2,:);
h3=hdyad(3,:);
y1=conv(x,h1);
y2=conv(x,h2);
y3=conv(x,h3);
ypoly=[y1;y2;y3];
ypoly=reshape(ypoly,1,size(ypoly,1)*size(ypoly,2));

%Magnitude response using the direct method and the poly-phase method
MAGx=[];
MAGz=[];
MAGy=[];
MAGPOLY=[];
n1=0:1:length(x)-1;
n2=0:1:length(z)-1;
n3=0:1:length(y)-1;
```

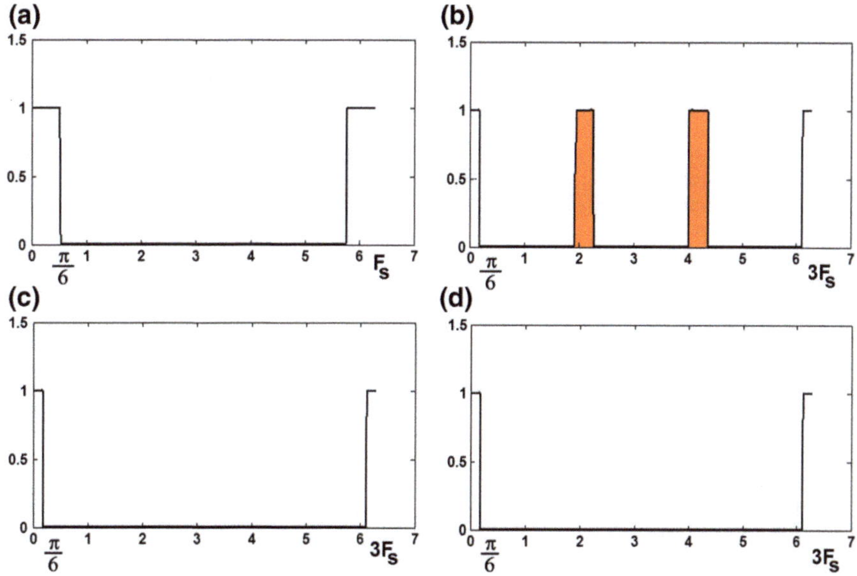

Fig. 4.6 Spectrum **a** Input signal **b** Signal after upsampling (inserting zeros) by 3 (*Shaded region* shows the mirror image). **c** Interpolated signal obtained by direct filtering using FIR filter with cutoff $\frac{\pi}{3}$. **d** Interpolated signal obtained using three poly-phase filters

```
n4=0:1:length(ypoly)-1;
for w=0:0.01:2*pi
s1=sum(x.*exp(-j*w*n1));
s2=sum(z.*exp(-j*w*n2));
s3=sum(y.*exp(-j*w*n3));
s4=sum(ypoly.*exp(-j*w*n4));
MAGx=[MAGx s1];
MAGz=[MAGz s2];
MAGy=[MAGy s3];
MAGPOLY=[MAGPOLY s4];
end


figure
subplot(2,2,1)
plot(linspace(0,2*pi,length(MAG)),abs(MAGx),'-')
subplot(2,2,2)
plot(linspace(0,2*pi,length(MAG)),abs(MAGz),'-')
subplot(2,2,3)
plot(linspace(0,2*pi,length(MAG)),abs(MAGy),'-')
subplot(2,2,4)
plot(linspace(0,2*pi,length(MAG)),abs(MAGPOLY),'-')

%Magnitude and Phase response of the efficient filter constructed
M=[];
P=[];
PPF1M=[];
```

```
PPF1P=[];
PPF2M=[];
PPF2P=[];
PPF3M=[];
PPF3P=[];
n=0:1:length(h)-1;
n1=0:1:length(h1)-1;
n2=0:1:length(h2)-1;
n3=0:1:length(h3)-1;
for w=0:0.01:2*pi
s=sum(h.*exp(-j*w*n));
s1=sum(h1.*exp(-j*w*n1));
s2=sum(h2.*exp(-j*w*n2));
s3=sum(h3.*exp(-j*w*n3));
M=[M abs(s)];
P=[P angle(s)];
PPF1M=[PPF1M abs(s1)];
PPF1P=[PPF1P angle(s1)];
PPF2M=[PPF2M abs(s2)];
PPF2P=[PPF2P angle(s2)];
PPF3M=[PPF3M abs(s3)];
PPF3P=[PPF3P angle(s3)];
end
figure
subplot(4,2,1)
plot(linspace(0,2*pi,length(M)), M)
subplot(4,2,3)
plot(linspace(0,2*pi,length(PPF1M)), PPF1M)
subplot(4,2,5)
plot(linspace(0,2*pi,length(PPF2M)), PPF2M)
subplot(4,2,7)
plot(linspace(0,2*pi,length(PPF3M)), PPF3M)
subplot(4,2,2)
plot(linspace(0,2*pi,length(P)), P)
subplot(4,2,4)
plot(linspace(0,2*pi,length(PPF1P)), PPF1P)
subplot(4,2,6)
plot(linspace(0,2*pi,length(PPF2P)), PPF2P)
subplot(4,2,8)
plot(linspace(0,2*pi,length(PPF3P)), PPF3P)
```

4.3 Poly-phase Realization of the Filter for Decimation

Decimation by the factor N is obtained by filtering the input sequence with the FIR filter with the cutoff $\frac{\pi}{N}$ and followed by down sampling the sequence (Removing $(M-1)$ samples between two consecutive samples). The corresponding poly-phase realization is as described below.

1. Consider the order of the filter (say 6) as the integer multiples of $M = 3$.
2. The impulse response of the filter is given as $h = [h_0\ h_1\ h_2\ h_3\ h_4\ h_5$. Note that the cutoff frequency is $\frac{pi}{M} = \frac{pi}{3}$. Obtain the three (in general M) polyfilters as $poly1 = [h_0\ h_3]$, $poly2 = [h_1\ h_4]$, and $poly3 = [h_2\ h_5]$.
3. It is observed that the $poly1$, $poly2$, and $poly3$ are the downsampled version of the filter h.
4. We have already identified that if the filter coefficients are downsampled, we get the average of shifted and scaled up version of the M original magnitude response (refer (1.11)). In particular if the cutoff frequency of the filter is $\frac{\pi}{M}$, we get the magnitude response as the flat (All pass).
5. Thus $poly1$, $poly2$, and $poly3$ are having flat magnitude response (All pass filter) (refer Fig. 4.5b–d)with different phase response. Hence the filter is named as poly-phase filter. The actual magnitude response of the filter is given in Fig. 4.6a.
6. The input sequence $[X_0\ X_1\ X_2\ X_3\ \cdots]$ is splitted into three sequences (In general N sequences) with the first sequence filled up with the elements collected as $S_1 = [X_0\ X_3\ X_6 \cdots]$. The second sequence is given as $S_2 = [X_1\ X_4\ X_5 \cdots]$ and the third sequence is given as $S_3 = [X_2\ X_5\ X_8 \cdots]$.
7. The sequence S_1 is filtered using poly-phase filter 1 namely $poly1$ to obtain out1. The sequence S_2 is filtered using poly-phase filter 3 namely $poly3$ to obtain out2. Similarly, the sequence S_3 is filtered using poly-phase filter 2 namely $poly2$ to obtain out2. The actual output is obtained as $out1 + out2 + out3$.
8. In general, the input sequence is divided into N sequences as described below. (a) Keeping first element of the input sequence as the first element of the new sequence and collecting one sample once in N samples gives the first sequence. (b) Keeping second element of the input sequence as the first element of the new sequence and collecting one sample once in N samples gives the second sequence. This is repeated to obtain N sequences (say $S_1, S_2, \cdots\ S_N$).
9. This method is applied to the impulse response of the decimator filter (with filter coefficient multiples of N) to obtain N poly-phase filter. (Say $poly1$, $poly_2$, $poly_3, \cdots, poly_{N-1}$ and $poly_N$).
10. The output sequences $out_1 = S_1 * poly_1, out_2 = S_2 * poly_N, out_3 = S_3 * poly_{N-1}$ $\cdots\ out_N = S_N * poly_2$ is computed. The final output sequence is obtained as $out = out_1 + out_2 + \cdots out_N$.
11. Trick to perform $\frac{M}{N}$ using poly-phase technique. (1) In the case $M > N$, choose the order of the filter as the multiples of M. Realize the poly-phase interpolation filter with cutoff frequency $\frac{\pi}{M}$ (refer Figs. 4.3, 4.4). This is followed by decimation by N. (2) In the case $M < N$, perform the interpolation by the factor N. Choose the order of the filter as the multiples of N with cutoff frequency $\frac{\pi}{N}$ and implement the poly-phase realization (refer Figs. 4.7, 4.8) of the decimation filter to obtain the final sequence (Figs. 4.9, 4.10, 4.11, 4.12).

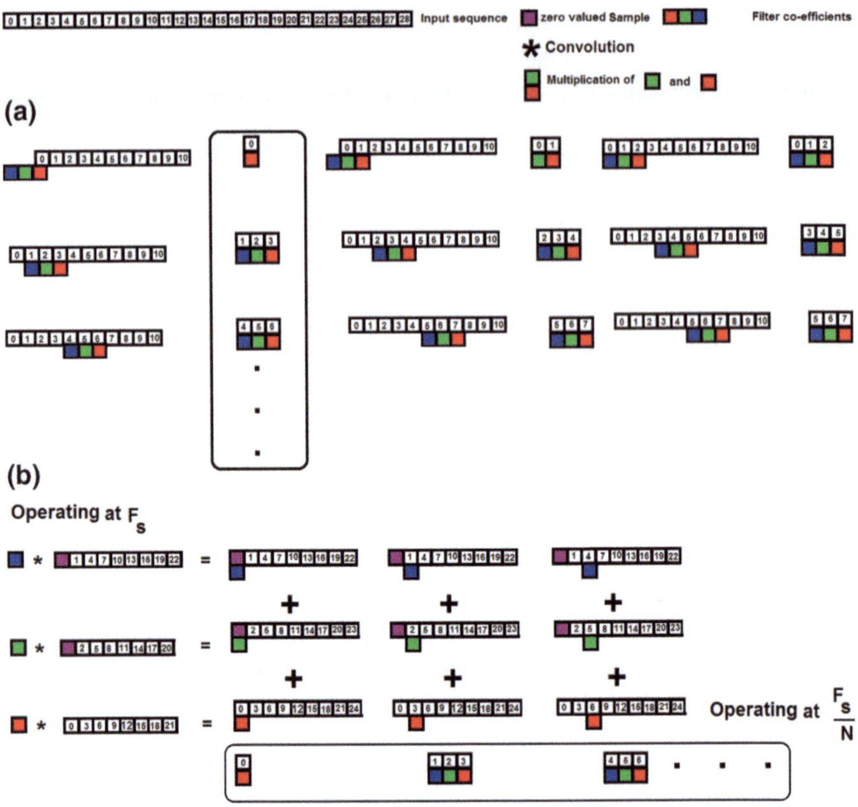

Fig. 4.7 Illustration of decimation by 3 using the decimator filter (order 3). **a** Direct method. **b** Poly phase method

```
%decimatepolyphase.m
%Decimation by 1/3
%Polyphase realization of sampling rate conversion
%by the factor 1/3
%Generating the input sequence
x=fir1(1000,3/4); %Not the filter coefficients.
MAG=[];
n=0:1:1000;
for w=0:0.01:2*pi
s=sum(x.*exp(-j*w*n));
MAG=[MAG s];
end

%Direct method (Decimation by the factor 3 )
x=x(1:1:999);
z=reshape(x,3,333);
z=z(1,:);
```

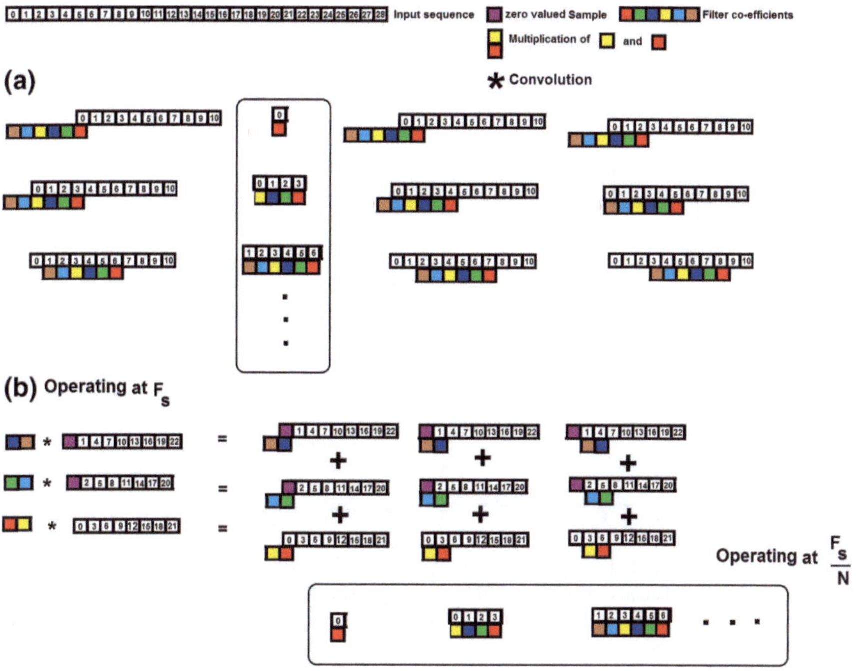

Fig. 4.8 Illustration of decimation by 3 using the decimator filter (order 6). **a** Direct method. **b** Poly-phase method

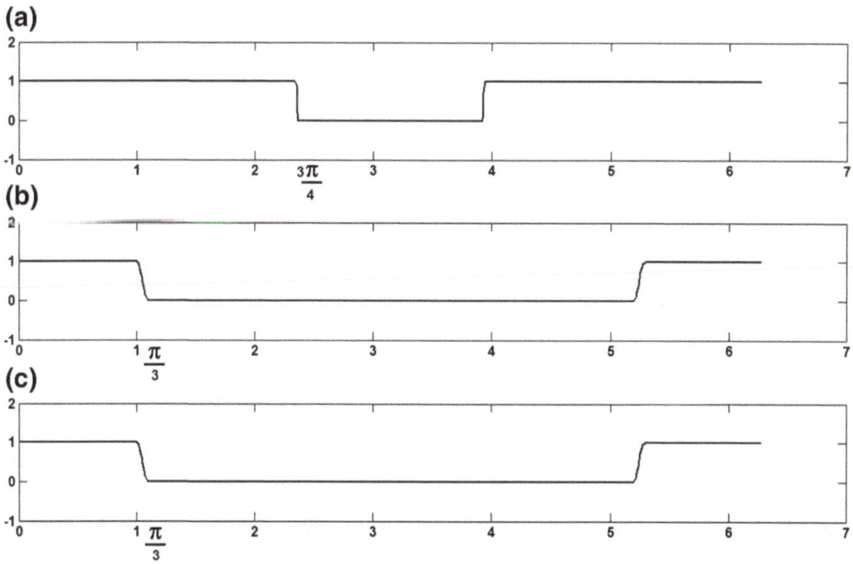

Fig. 4.9 **a** Spectrum of the input signal with the maximum frequency content as $\frac{3\pi}{4}$. **b** Filter used before decimation with cutoff frequency $\frac{\pi}{3}$. **c** Spectrum of the output signal after filtering before subjected to decimation

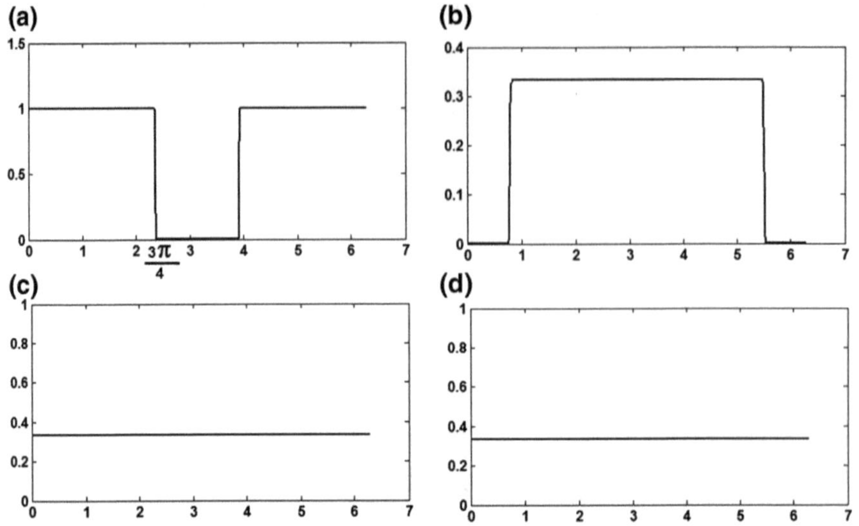

Fig. 4.10 **a** Spectrum of the input signal. **b** Spectrum of the decimated signal without filtering (overlapping occurs because maximum frequence $\frac{3\pi}{4}$ exceeds $\frac{\pi}{3}$). **c** Spectrum of the decimated signal after filtering using the direct method. **d** Spectrum of the decimated signal after filtering using poly-phase method

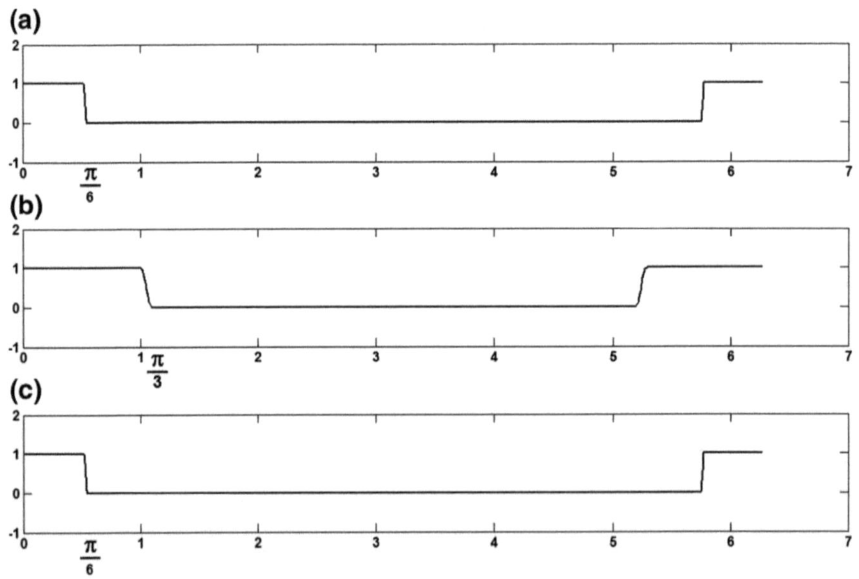

Fig. 4.11 **a** Spectrum of the input signal with the maximum frequency content as $\frac{\pi}{6}$. **b** Filter used before decimation with cutoff frequency $\frac{\pi}{3}$. **c** Spectrum of the output signal after filtering before subjected to decimation (Identical with the spectrum of the input signal)

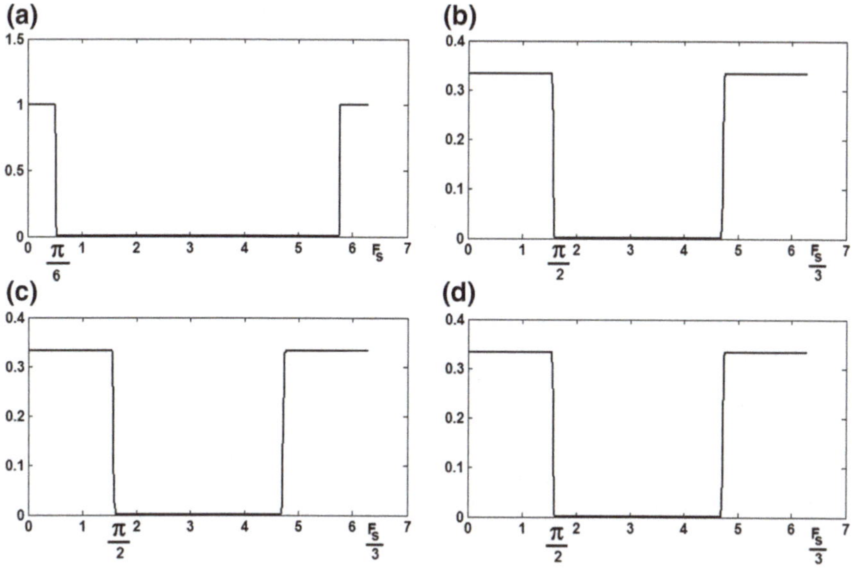

Fig. 4.12 **a** Spectrum of the input signal. **b** Spectrum of the decimated signal without filtering (overlapping does not occurs because maximum frequency $\frac{\pi}{6}$ does not exceed $\frac{\pi}{3}$). **c** Spectrum of the decimated signal after filtering using the direct method. **d** Spectrum of the decimated signal after filtering using poly-phase method

Fig. 4.13 Block diagram. **a** Quadrature mirror filter. **b** transmultiplexer

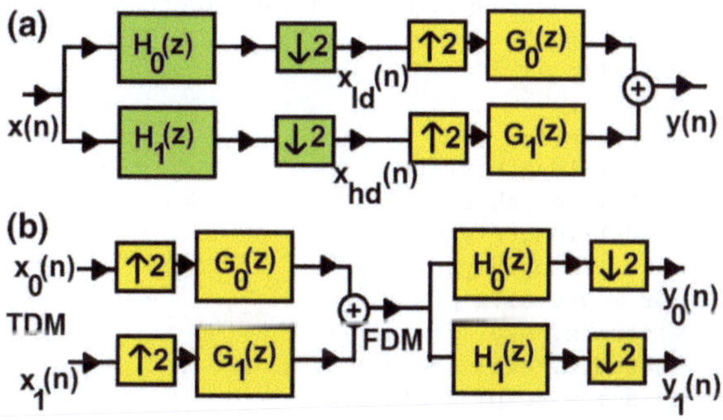


```
h=fir1(212,1/3);
y1=conv(x,h);
MAGx=[];
MAGh=[];
MAGy1=[];
n1=0:1:length(x)-1;
n2=0:1:length(h)-1;
n3=0:1:length(y1)-1;
for w=0:0.01:2*pi
s1=sum(x.*exp(-j*w*n1));
s2=sum(h.*exp(-j*w*n2));
s3=sum(y1.*exp(-j*w*n3));
```

```
MAGx=[MAGx s1];
MAGh=[MAGh s2];
MAGy1=[MAGy1 s3];
end
figure
subplot(3,1,1)
plot(0:0.01:2*pi,abs(MAGx))
subplot(3,1,2)
plot(0:0.01:2*pi,abs(MAGh))
subplot(3,1,3)
plot(0:0.01:2*pi,abs(MAGy1))

y=y1(1:1:1209);
y=reshape(y,3,403);
y=y(1,:);
x=[0 0 x];
x=x(1:1:999);
x1=reshape(x,3,333);
%Realization using poly-phase filter.
hdyad=reshape(h,3,length(h)/3);
h1=hdyad(1,:);
h2=hdyad(2,:);
h3=hdyad(3,:);
y1=conv(x1(1,:),h3);
y2=conv(x1(2,:),h2);
y3=conv(x1(3,:),h1);
ypoly=y1+y2+y3;

%Magnitude response using the direct method and the poly-phase method
MAGx=[];
MAGz=[];
MAGy=[];
MAGPOLY=[];
n1=0:1:length(x)-1;
n2=0:1:length(z)-1;
n3=0:1:length(y)-1;
n4=0:1:length(ypoly)-1;
for w=0:0.01:2*pi
s1=sum(x.*exp(-j*w*n1));
s2=sum(z.*exp(-j*w*n2));
s3=sum(y.*exp(-j*w*n3));
s4=sum(ypoly.*exp(-j*w*n4));
MAGx=[MAGx s1];
MAGz=[MAGz s2];
MAGy=[MAGy s3];
MAGPOLY=[MAGPOLY s4];
end

figure
subplot(2,2,1)
plot(linspace(0,2*pi,length(MAG)),abs(MAGx),'-')
subplot(2,2,2)
plot(linspace(0,2*pi,length(MAG)),abs(MAGz),'-')
subplot(2,2,3)
plot(linspace(0,2*pi,length(MAG)),abs(MAGy),'-')
subplot(2,2,4)
```

```
plot(linspace(0,2*pi,length(MAG)),abs(MAGPOLY),'-')

%Magnitude and Phase response of the efficient filter constructed
M=[];
P=[];
PPF1M=[];
PPF1P=[];
PPF2M=[];
PPF2P=[];
PPF3M=[];
PPF3P=[];
n=0:1:length(h)-1;
n1=0:1:length(h1)-1;
n2=0:1:length(h2)-1;
n3=0:1:length(h3)-1;
for w=0:0.01:2*pi
s=sum(h.*exp(-j*w*n));
s1=sum(h1.*exp(-j*w*n1));
s2=sum(h2.*exp(-j*w*n2));
s3=sum(h3.*exp(-j*w*n3));
M=[M abs(s)];
P=[P angle(s)];
PPF1M=[PPF1M abs(s1)];
PPF1P=[PPF1P angle(s1)];
PPF2M=[PPF2M abs(s2)];
PPF2P=[PPF2P angle(s2)];
PPF3M=[PPF3M abs(s3)];
PPF3P=[PPF3P angle(s3)];
end
figure
subplot(4,2,1)
plot(linspace(0,2*pi,length(M)), M)
subplot(4,2,3)
plot(linspace(0,2*pi,length(PPF1M)), PPF1M)
subplot(4,2,5)
plot(linspace(0,2*pi,length(PPF2M)), PPF2M)
subplot(4,2,7)
plot(linspace(0,2*pi,length(PPF3M)), PPF3M)
subplot(4,2,2)
plot(linspace(0,2*pi,length(P)), P)
subplot(4,2,4)
plot(linspace(0,2*pi,length(PPF1P)), PPF1P)
subplot(4,2,6)
plot(linspace(0,2*pi,length(PPF2P)), PPF2P)
subplot(4,2,8)
plot(linspace(0,2*pi,length(PPF3P)), PPF3P)
```

4.4 Quadrature Mirror Filter

Quadrature Mirror Filter consists of L-channel analysis filter bank at the input and L-channel synthesis filter at the output end (refer Fig. 4.13a). Consider $L = 2$. Let the transfer function of the two-channel QMF (analysis (H) and synthesis (G)) filter are as shown below:

$$H_0(z) = \frac{1}{2}(A_0(z^2) + z^{-1}A_1(z^2))$$
$$H_1(z) = \frac{1}{2}(A_0(z^2) - z^{-1}A_1(z^2))$$
$$G_0(z) = \frac{1}{2}(A_0(z^2) + z^{-1}A_1(z^2))$$
$$G_1(z) = \frac{1}{2}(z^{-1}A_1(z^2) - A_0(z^2)).$$

where $A_0(z)$ and $A_1(z)$ are the all pass filter. Let the z-transformation of the signal after analysis filter be represented as $U_0(z)$, $U_1(z)$ and are computed as follows:

$$U_0(z) = \frac{1}{2}(H_0(z^{1/2}) + H_0(-z^{1/2}))$$
$$U_0(z) = \frac{1}{4}(A_0(z) + z^{-1/2}A_1(z) + A_0(z) - z^{-1/2}A_1(z) = \frac{1}{2}A_0(z))$$
$$U_1(z) = \frac{1}{2}(H_1(z^{1/2}) + H_1(-z^{1/2}))$$
$$U_1(z) = \frac{1}{4}(A_0(z) - z^{-1/2}A_1(z) + A_0(z) + z^{-1/2}A_1(z)) = \frac{1}{2}A_0(z).$$

It is upsampled before given to the synthesis filter and hence the overall transfer function is given as

$$U_0(z^2)G_0(z) + U_1(z^2)G_1(z)$$
$$= \frac{1}{4}A_0(z^2)G_0(z) + \frac{1}{4}A_0(z^2)G_1(z)$$
$$= \frac{1}{8}(A_0(z^2) + z^{-1}A_1(z^2))A_0(z^2) + \frac{1}{8}(z^{-1}A_1(z^2) - A_0(z^2))A_0(z^2)$$
$$= \frac{1}{4}A_0(z^2)A_1(z^2)z^{-1}.$$

The typical analysis filters and the synthesis filters are constructed using the all pass filters as given in (4.12 and 4.13). Figure 4.14 shows the magnitude and phase response of the constructed filters. The speech signal is decomposed using the constructed analysis filters and are reconstructed using the typical synthesis filters are shown in Fig. 4.15.

$$A_0(z) = \frac{0.10557281 + z^{-2}}{1 + 0.1055781z^{-2}} \tag{4.12}$$

$$A_1(z) = \frac{0.527864045 + z^{-2}}{1 + 0.527864045z^{-2}} \tag{4.13}$$

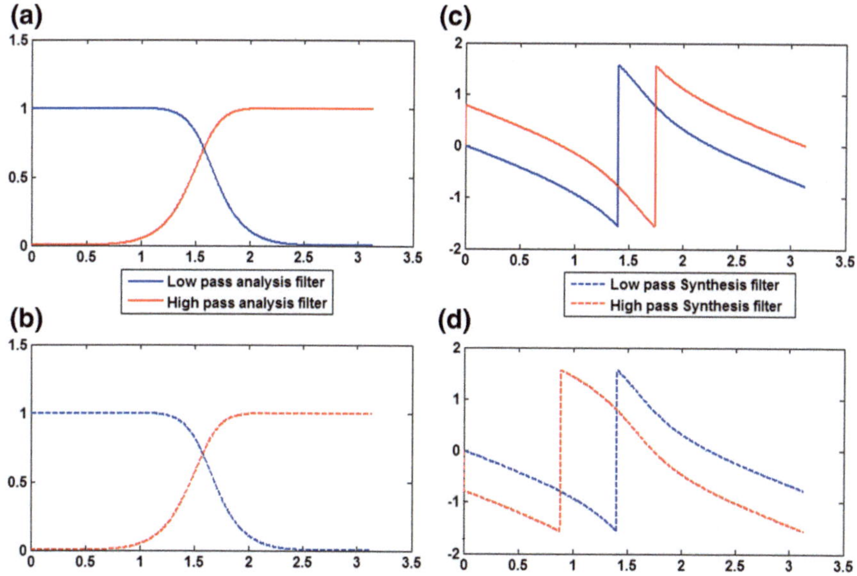

Fig. 4.14 Typical QMF filters. **a** Magnitude response of $H_0(z)$ and $H_1(z)$. **b** Phase response of $H_0(z)$ and $H_1(z)$. **c** Magnitude response of $G_0(z)$ and $G_1(z)$. **d** Phase response of $G_0(z)$ and $G_1(z)$

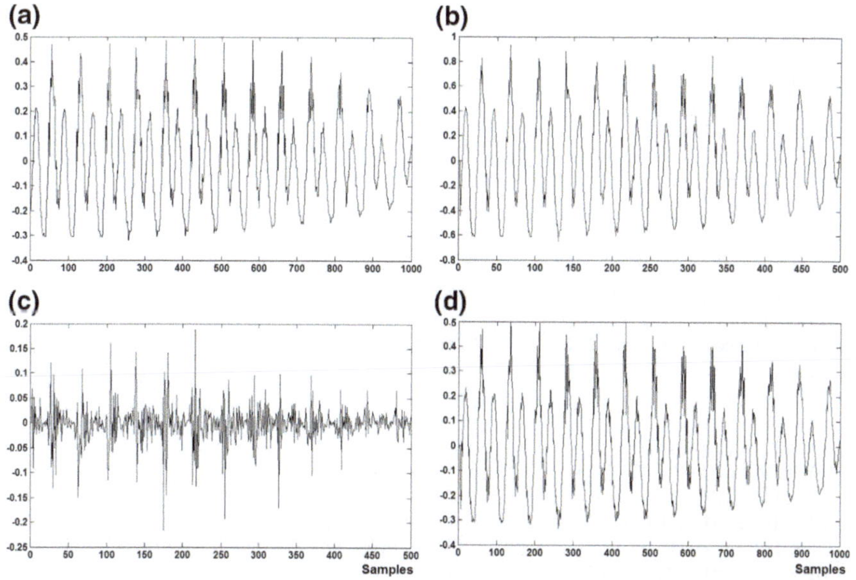

Fig. 4.15 **a** Input speech signal. **b** Output of low-pass decomposition filter. **c** Output of high pass decomposition filter. **d** Signal reconstructed with low pass and high-pass reconstruction filter

```
%QMFdemo.m
clear all
close all
a0d=0.10557281;
a1d=0.10557281;
b0d=0.527864045;
b1d=0.527864045;
a0r=0.10557281;
a1r=0.10557281;
b0r=0.527864045;
b1r=0.527864045;
i=2;
QMF
part1=recons;
save part1 part1
clear all
a0d=0.10557281;
a1d=0.10557281;
b0d=-0.527864045;
b1d=0.527864045;
a0r=-0.10557281;
a1r=0.10557281;
b0r=0.527864045;
b1r=0.527864045;
i=3;
QMF
part2=recons;
load part1
part=(part1+part2)/2;
subplot(2,2,1)
plot(speechdata)
subplot(2,2,4)
plot(part)
coef1=a0d;
coef2=b1d;


%A0
[H1,W]=freqz([coef1 0 1],[1 0 coef1]);
%z-1A1
[H2,W]=freqz([0 coef2 0 1],[1 0 coef2]);
figure
subplot(2,2,1)
plot(W,abs(H1+H2)/2)
subplot(2,2,2)
plot(W,angle(H1+H2)/2)
subplot(2,2,3)
plot(W,abs(H1+H2)/2)
subplot(2,2,4)
plot(W,angle(H1+H2)/2)
%-z-1A1
[H3,W]=freqz([-0 -coef2 -0 -1],[1 0 coef2]);
subplot(2,2,1)
hold on
plot(W,abs(H1+H3)/2,'r')
subplot(2,2,2)
hold on
```

```
plot(W,angle(H1+H3)/2,'r')
%-A0
[H4,W]=freqz([-coef1 -0 -1],[1 0 coef1]);
subplot(2,2,3)
hold on
plot(W,abs(H2+H4)/2,'r')
subplot(2,2,4)
hold on
plot(W,angle(H2+H4)/2,'r')


%QMF.m
%Called by the script QMFdemo.m
load SPEECHDATA
%A0(0.10557281+z^{-2})/(1+0.10557281z^{-2});
%A1=(0.527864045+z^{-2})/(1+0.527864045z^{-2});
%Decomposition
speechdata=speechdata(2000:1:3000);
x=speechdata;
y=zeros(1,3);
for n=4:1:length(speechdata)
    y(n)=a0d*x(n)+sign(a1d)*x(n-2)-a1d*y(n-2);
end
output1=y;

x=speechdata;
y=zeros(1,3);
for n=4:1:length(x)
    y(n)=b0d*x(n-1)+sign(b0d)*x(n-3)-b1d*y(n-2);
end
output2=y;
z=output1+output2;
%Downsampling
L=fix(length(z)/2);
z=z(1:1:2*L);
output=reshape(z,2,L);
output=output(1,:);
subplot(2,2,i)
plot(output)
%Upsampling
output=[output;zeros(1,length(output))];
p=reshape(output,1,size(output,1)*size(output,2));
%%%%%%%%%%%%%%%%%%%%%%%%%%%%%%%%%%%%%%%%%%%%%%%%%%
%Reconstruction
x=p;
y=zeros(1,3);
for n=4:1:length(x)
    y(n)=a0r*x(n)+sign(a0r)*x(n-2)-a1r*y(n-2);
end
output1=y;
x=p;
y=zeros(1,3);
for n=4:1:length(x)
    y(n)=b0r*x(n-1)+sign(b0r)*x(n-3)-b1r*y(n-2);
end
output2=y;
recons=output1+output2;
```

4.5 Transmultiplexer

Time division multiplexing involves allotting alternative time slots to the individual sequences to share the channel. Frequency division multiplexing (FDM) is the method of allotting different bandwidths for the individual sequences to share the common channel. Trans-multiplexer (refer Fig. 4.13b) helps to get the new sequence from the actual two sequences to be transmitted through the channel that converts TDM to FDM sharing of the channel.

Consider two sequences in Fig. 4.16a, b to be transmitted through the channel. The spectrum of these two sequences is given in Fig. 4.17a, b, respectively. We obtain the new sequence as shown in Fig. 4.16c whose spectrum is given in Fig. 4.19c. In this case the spectrum occupied by the individual sequence are identified at different slots. This is achieved by interpolating the two sequences individually and followed by the corresponding decomposition filters (Fig. 4.18a, b). The spectrum of the sequence 1 after subjected to interpolation, followed by low-pass decomposition filter is given in Fig. 4.19a. Similarly, the spectrum of the sequence 2 after subjected to interpolation,followed by high-pass decomposition filter is given in Fig. 4.19b. This shows how the spectrum of the channel is shared between the two sequences. The received sequence (assuming ideal channel), is passed through the low-pass reconstruction filter Fig. 4.18c, d, followed by down sampling to obtain the sequence 1. Similarly, the received sequence is passed through the high-pass reconstruction filter, followed by down sampling to obtain the sequence 2.

```
%transmultiplexer.m
X00=fir2(100,linspace(0,1,1000),triang(1000)');
X11=fir2(100,linspace(0,1,1000),(1-triang(1000))');

%G0z=z^{-1}+z^{-2};
%G1z=z^{-1}-z^{-2};
%H0z=1+z^{-1};
%H1z=1-z^{-1};
%TDM2FDM
X0=[X00;zeros(1,length(X00))];
X0=reshape(X0,1,size(X0,1)*size(X0,2));
X1=[X11;zeros(1,length(X11))];
X1=reshape(X1,1,size(X1,1)*size(X1,2));
X0=[zeros(1,2) X0];
X1=[zeros(1,2) X1];
for n=3:1:length(X0)
Y0(n)= X0(n-1)+X0(n-2);
Y1(n)=X1(n-1)-X1(n-2);
end
Y=Y0+Y1;
n=0:1:length(Y0)-1;
m0=[];
m1=[];
m=[];
for w=-pi:0.001:pi;
    m0=[m0 abs(sum(Y0.*exp(-j*w*n)))];
    m1=[m1 abs(sum(Y1.*exp(-j*w*n)))];
```

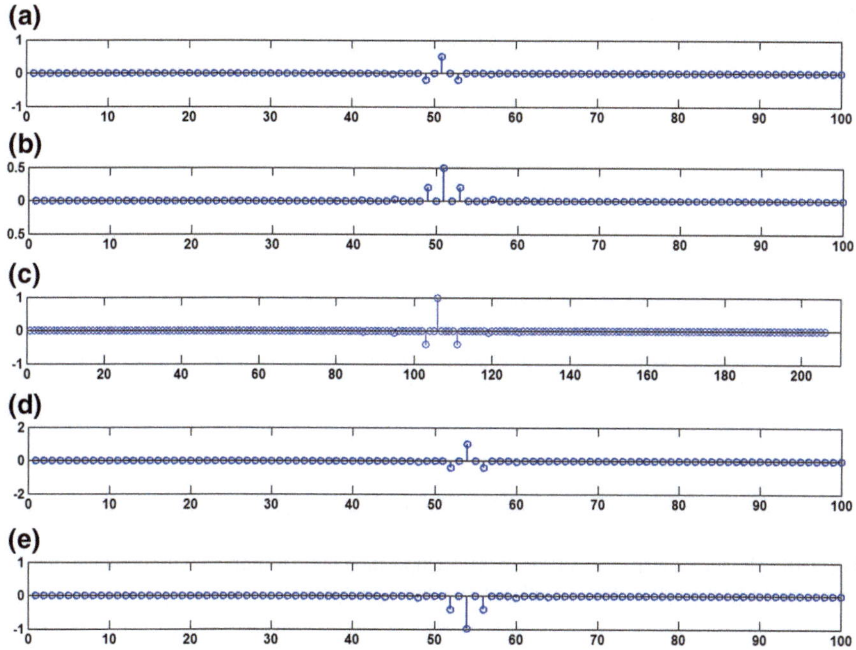

Fig. 4.16 **a** Sequence 1 in the transmitter. **b** Sequence 2 in the transmitter. **c** Sequence transmitted through the channel. **d** Sequence 1 in the receiver. **e** Sequence 2 in the receiver

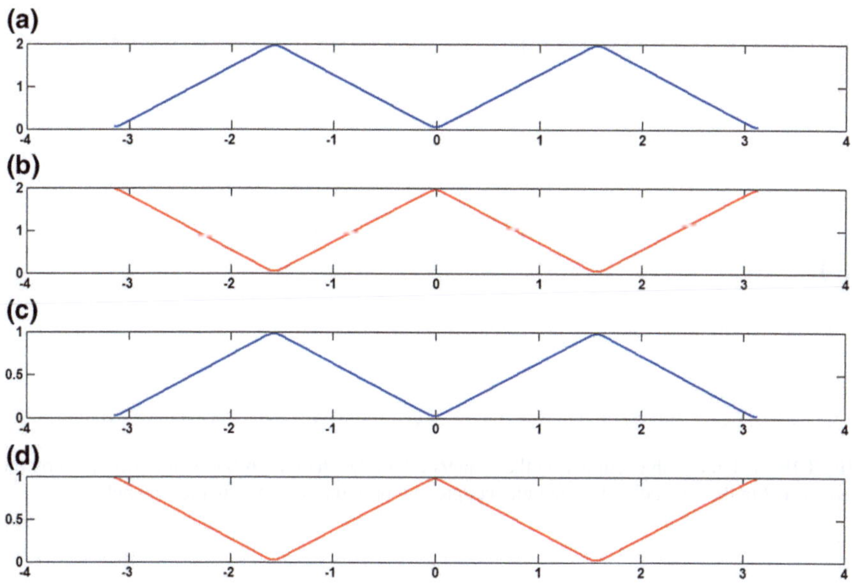

Fig. 4.17 **a** Actual spectrum occupied by the sequence 1 before transmission. **b** Actual spectrum occupied by the sequence 2 before transmission. **c** Actual spectrum occupied by the sequence 1 in the receiver. **d** Actual spectrum occupied by the sequence 2 in the receiver

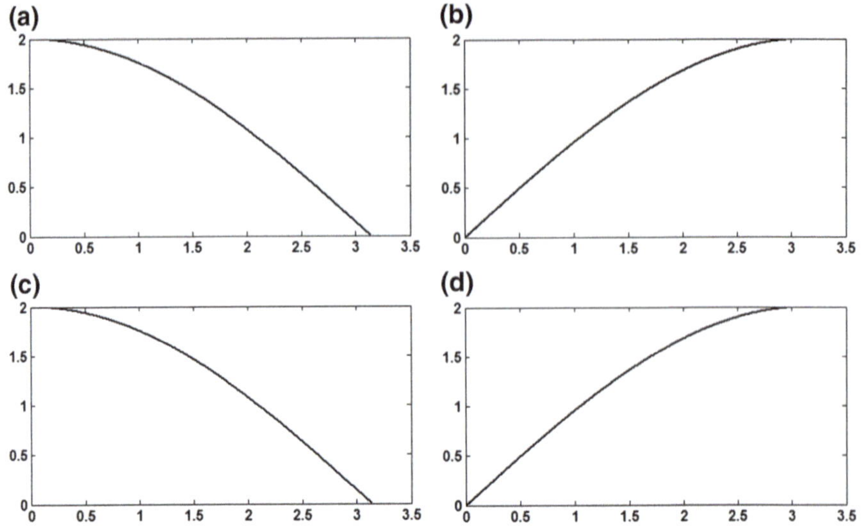

Fig. 4.18 Magnitude response: **a** Low-pass decomposition filter. **b** High-pass decomposition filter. **c** Low-pass reconstruction filer. **d** High-pass reconstruction filter

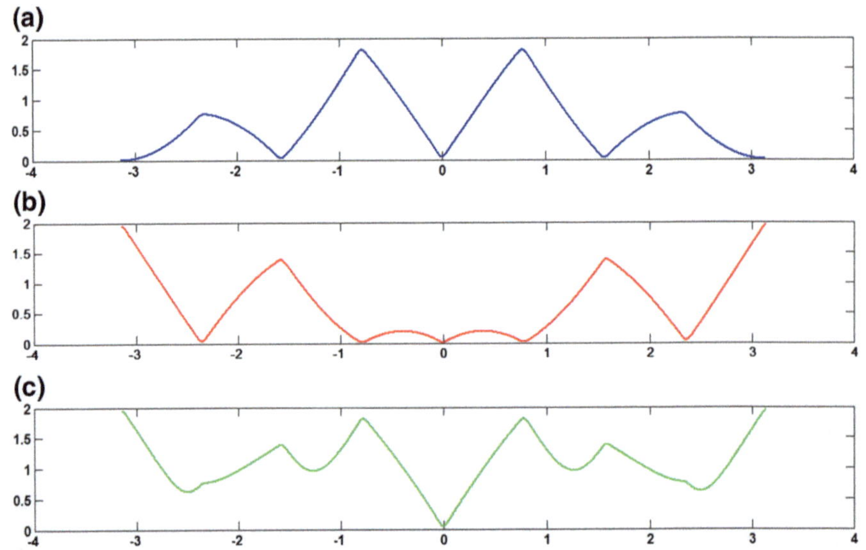

Fig. 4.19 **a** Bandwidth occupied by the sequence 1 in the channel. **b** Bandwidth occupied by the sequence 2 in the channel. **c** Bandwidth occupied by both the sequence in the channel

```
    m=[m abs(sum(Y.*exp(-j*w*n)))];
end
figure
subplot(3,1,1)
plot(-pi:0.001:pi,m0)
subplot(3,1,2)
plot(-pi:0.001:pi,m1,'r')
subplot(3,1,3)
plot(-pi:0.001:pi,m,'g')

%FDM2TDM
Y=[zeros(1,2) Y];
for n=3:1:length(Y)
XR0(n)=Y(n)+Y(n-1);
XR1(n)=Y(n)-Y(n-1);
end
XR00=dyaddown(XR0,1);
XR11=dyaddown(XR1,1);

m0=[];
m1=[];
m2=[];
m3=[];
n1=0:1:length(XR00)-1;
n2=0:1:length(X00)-1;

for w=-pi:0.001:pi;
    m0=[m0 abs(sum(XR00.*exp(-j*w*n1)))];
    m1=[m1 abs(sum(XR11.*exp(-j*w*n1)))];
    m2=[m2 abs(sum(X00.*exp(-j*w*n2)))];
    m3=[m3 abs(sum(X11.*exp(-j*w*n2)))];
end

figure
subplot(4,1,1)
plot(-pi:0.001:pi,m0)
subplot(4,1,2)
plot(-pi:0.001:pi,m1,'r')
subplot(4,1,3)
plot(-pi:0.001:pi,m2)
subplot(4,1,4)
plot(-pi:0.001:pi,m3,'r')


figure
subplot(5,1,1)
stem(X00)
subplot(5,1,2)
stem(X11)
subplot(5,1,3)
stem(Y)
subplot(5,1,4)
stem(XR00)
subplot(5,1,5)
stem(XR11)
```

```
figure
[H1,W1]=freqz([0 1 1],1);
subplot(2,2,1)
plot(W1,abs(H1))
[H2,W2]=freqz([0 1 -1],1);
subplot(2,2,2)
plot(W2,abs(H2))
[H3,W3]=freqz([1 1],1);
subplot(2,2,3)
plot(W1,abs(H1))
[H4,W4]=freqz([1 -1],1);
subplot(2,2,4)
plot(W2,abs(H2))
```

Chapter 5
Statistical Signal Processing

5.1 Introduction to Random Process

The outcome of the experiments is mapped to the function of time and is known as random process. Random process is analyzed by obtaining the random variable X_t tapped across the process at some time instant t. The random process is said to be strictly stationary random process (S.S.R.P) with first order if $FX_t = FX_{t+\tau}$ for all t and τ.

The random process is said to be strictly stationary random process with second order if $FX_{t1,t2} = FX_{t1+\tau,t2+\tau}$ for all t_1, t_2 and τ.

In practice, identifying whether the given random process is S.S.R.P is difficult. Hence we identify the particular set of random process known as wide sense stationary random process (W.S.S.R.P), which satisfies the following conditions:

1. $E(X_t)$ = constant
2. $E(X_{t+\tau}X_t) = R_X(\tau)$.

5.1.1 *Illustration on W.S.S.R.P and Nonstationary Random Process*

The discrete random process X_n is generated with the outcome of X_n satisfies the following conditions:

1. The outcome of X_n takes the value 1, 2, 3 or 4
2. X_n is identically distributed for all n
3. X_{n1} and X_{n2} are independent for all n_1 and n_2.

E.S. Gopi, *Multi-Disciplinary Digital Signal Processing*,
DOI 10.1007/978-3-319-57430-1_5

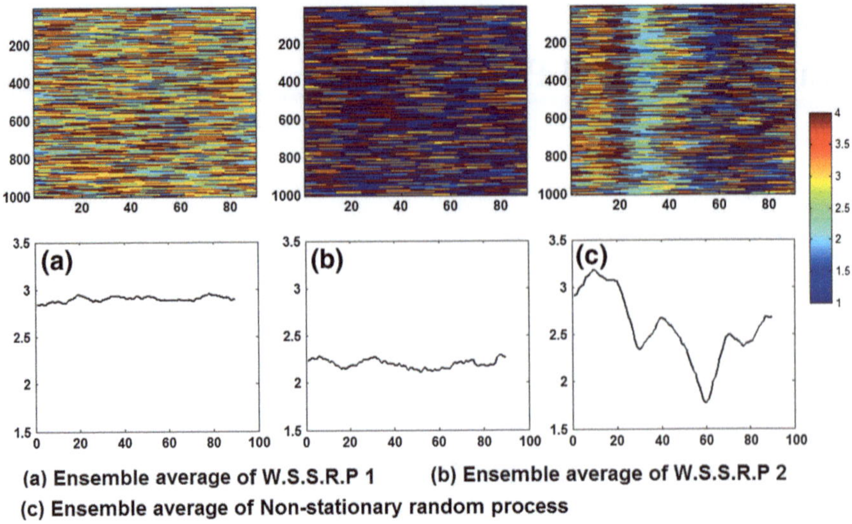

Fig. 5.1 Illustration of random process. **a** W.S.S.R.P 1. **b** W.S.S.R.P 2. **c** N.S.R.P

Hundred outcomes are stacked row wise to form the image and are displayed in the first row of Fig. 5.1a, b. The corresponding estimated mean across the process for every n is plotted in the Fig. 5.1a, b (second row), respectively. Further, the experiment is repeated with nonuniform distribution for all n. Fig. 5.1c (first row and second row) shows the image representation of the random process and the corresponding estimated mean. Also the sample outcome of the random process under each cases is displayed in the Fig. 5.2. It is seen from Fig. 5.1, the estimated mean is almost constant in (a) and (b) and have large variation in (c). This demonstrates the W.S.S.R.P (subplot a and b) and nonstationary random process (subplot c).

```
%wssrp.m
clear all
PR=rand(1,4);
PR=PR/sum(PR);
R=cumsum(PR);
DATA=[];
for i=1:1:100
D=[];
for j=1:1:1000
P=R-rand;
[U,V]=find(P>0);
D=[D;V(1)];
end
DATA=[DATA repmat(D,1,10)];
end
FINALDATA=[];
for i=1:1:1000
    t=round(rand*9);
    FINALDATA=[FINALDATA ;...
zeros(1,t) DATA(i,:) zeros(1,10-t)];
```

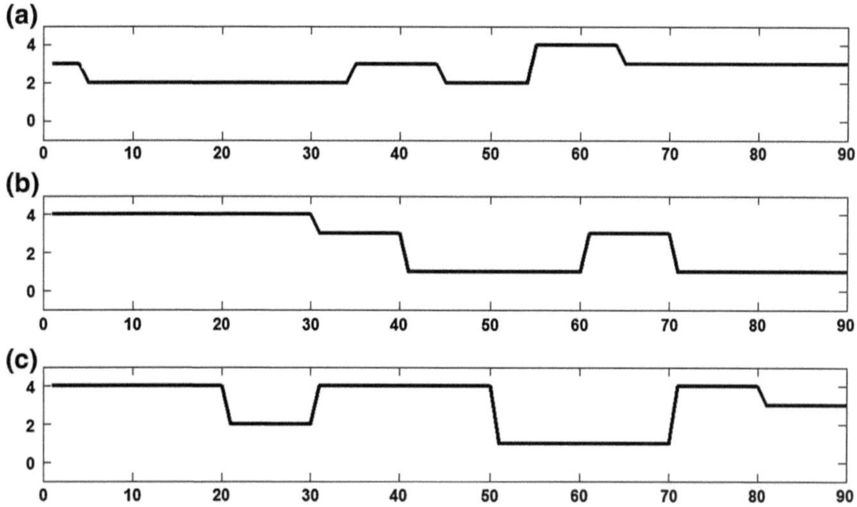

Fig. 5.2 **a** 500th row of W.S.S.R.P 1. **b** 500th row of W.S.S.R.P 2. **c** 500th row of N.S.R.P

```
end
temp=FINALDATA(:,11:1:100);
figure(1)
subplot(2,3,1)
imagesc(temp)
subplot(2,3,4)
plot(mean(temp))
figure(2)
subplot(3,1,1)
plot(temp(500,:))

clear all
PR=rand(1,4);
PR=PR/sum(PR);
R=cumsum(PR);
DATA=[];
for i=1:1:100
D=[];
for j=1:1:1000
P=R-rand;
[U,V]=find(P>0);
D=[D;V(1)];
end
DATA=[DATA repmat(D,1,10)];
end
FINALDATA=[];
for i=1:1:1000
    t=round(rand*9);
    FINALDATA=[FINALDATA ;...
zeros(1,t) DATA(i,:) zeros(1,10-t)];
end
temp=FINALDATA(:,11:1:100);
figure(1)
```

```
subplot(2,3,2)
imagesc(temp)
subplot(2,3,5)
plot(mean(temp))
figure(2)
subplot(3,1,2)
plot(temp(500,:))
clear all
DATA=[];
for i=1:1:100
D=[];
PR=rand(1,4);
PR=PR/sum(PR);
R=cumsum(PR);
for j=1:1:1000
P=R-rand;
[U,V]=find(P>0);
D=[D;V(1)];
end
DATA=[DATA repmat(D,1,10)];
end
FINALDATA=[];
for i=1:1:1000
    t=round(rand*9);
    FINALDATA=[FINALDATA ;...
zeros(1,t) DATA(i,:) zeros(1,10-t)];
end
temp=FINALDATA(:,11:1:100);
figure(1)
subplot(2,3,3)
    imagesc(temp)
subplot(2,3,6)
plot(mean(temp))
figure(2)
subplot(3,1,3)
plot(temp(500,:))
```

5.2 Auto Regressive (AR), Moving Average (MA) and Auto Regressive Moving Average (ARMA) Modeling

The W.S.S. random process W_n is said to be white if it satisfies the following:

$$\gamma_\tau = 1 for \tau = 0 = 0, \quad otherwise \tag{5.1}$$

and the spectral density is flat as given below:

$$S_w(f) = 1 \tag{5.2}$$

for all f.

Let the Wide Sense Stationary (WSS) random process, which is white W_n (with variance σ_w^2) is given as the input to the system $H(z)$ and the spectral density of the corresponding output is given as $S_x(z) = \sigma_w^2 H(z)H(1/z)$ with $r_1 < |z| < r_2$. Given the spectral density of the W.S.S. random process X_n, obtaining $H(z)$ that generates the outcome of random process X_n with the input W.S.S. white random process W_n with the identified variance σ_w^2 is known as wold representation. This is used to model the generation of typical W.S.S. random process (such as speech signal) with the specific spectral density. Let $H(z)$ is represented as follows:

$$H(z) = \frac{b_0 + b_1 z^{-1} + b_2 z^{-2} + b_3 z^{-3}}{1 + a_1 z^{-1} + a_2 z^{-2} + a_3 z^{-3}} \tag{5.3}$$

The input W.S.S. random process W_n and the output W.S.S. random process X_n are related as follows.

$$X_n = b_0 W_n + b_1 W_{n-1} + b_2 W_{n-2} + b_3 W_{N-3} - a_1 X_{n-1} - a_2 X_{n-2} - a_3 X_{n-3}. \tag{5.4}$$

This is known as ARMA (Auto regressive, Moving Average model). If we choose $a_1 = a_2 = 0$, the model is known as MA (Moving Average model). If we choose $b_1 = b_2 = 0$ and $b_0 = 1$, the model is known as AR (Auto regressive model). The coefficients are obtained as follows.

5.2.1 Linear Prediction Model

We see the expression $X_n = a_1 X_{n-1} + a_2 X_{n-2} + a_3 X_{n-3}$ as the prediction of the nth sample of the random process X_n using the previous samples. This is known as forward prediction. It is also observed that $W_n = X_n - a_1 X_{n-1} - a_2 X_{n-2} - a_3 X_{n-3} =$ is the error involved in predicting the nth. Thus the coefficients are obtained by minimizing the mean squared error $E((W_n)^2)$. (Note that we are considering only the real signals.) We obtain the coefficients such that $E((X_n + a_1 X_{n-1} + a_2 X_{n-2} + a_3 X_{n-3})^2)$ is minimized. Differentiating with respect to a_1, a_2, a_3 and equate to zero, we get the following:

$$\begin{aligned}
E(-2(X_n - a_1 X_{n-1} - a_2 X_{n-2} - a_3 X_{n-3})(X_{n-1})) = 0 \\
\Rightarrow \gamma_x(1) = a_1 \gamma_x(0) a_2 \gamma_x(-1) a_3 \gamma_x(-2) \\
E(-2(X_n - a_1 X_{n-1} - a_2 X_{n-2} - a_3 X_{n-3})(X_{n-2})) = 0 \\
\Rightarrow \gamma_x(2) = a_1 \gamma_x(1) a_2 \gamma_x(0) a_3 \gamma_x(-1) \\
E(-2(X_n - a_1 X_{n-1} - a_2 X_{n-2} - a_3 X_{n-3})(X_{n-3})) = 0 \\
\Rightarrow \gamma_x(2) = a_1 \gamma_x(1) a_2 \gamma_x(0) a_3 \gamma_x(0)
\end{aligned}$$

Rewriting in the matrix form we get the following:

$$\begin{bmatrix} \gamma_x(0) & \gamma_x(-1) & \gamma_x(-2) \\ \gamma_x(1) & \gamma_x(0) & \gamma_x(-1) \\ \gamma_x(2) & \gamma_x(1) & \gamma_x(0) \end{bmatrix} \begin{bmatrix} a_1 \\ a_2 \\ a_3 \end{bmatrix} = \begin{bmatrix} \gamma_x(1) \\ \gamma_x(2) \\ \gamma_x(3) \end{bmatrix}$$

It is noted that a_1, a_2 and a_3 are AR coefficients. Suppose that we represent $b_0 X_n + b_1 X_{n-1} + b_2 X_{n-2}$ as the prediction of the X_{n-3} of the random process X_n. We would like to obtain the coefficients b_0, b_1 and b_2 that minimizes.

$$E((X_{n-3} - b_0 X_n - b_1 X_{n-1} - b_2 X_{n-2})^2) \tag{5.5}$$

Differentiating (5.5) with respect to b_0, b_1 and b_2 and equate to zero, we get the following:

$$E(-2(X_{n-3} - b_0 X_n - b_1 X_{n-1} - b_2 X_{n-2})X(n)) = 0 \tag{5.6}$$
$$\Rightarrow \gamma_x(-3) = b_0\gamma_x(0) + b_1\gamma_x(-1) + b_2\gamma_x(-2) = 0 \tag{5.7}$$
$$E(-2(X_{n-3} - b_0 X_n - b_1 X_{n-1} - b_2 X_{n-2})X(n-1)) = 0 \tag{5.8}$$
$$\Rightarrow \gamma_x(-2) = b_0\gamma_x(1) + b_1\gamma_x(0) + b_2\gamma_x(-1) = 0 \tag{5.9}$$
$$E(-2(X_{n-3} - b_0 X_n - b_1 X_{n-1} - b_2 X_{n-2})X(n-2)) = 0 \tag{5.10}$$
$$\Rightarrow \gamma_x(-1) = b_0\gamma_x(2) + b_1\gamma_x(1) + b_2\gamma_x(0) = 0 \tag{5.11}$$

Representing in the matrix form, we get the following:

$$\begin{bmatrix} \gamma_x(0) & \gamma_x(-1) & \gamma_x(-2) \\ \gamma_x(1) & \gamma_x(0) & \gamma_x(-1) \\ \gamma_x(2) & \gamma_x(1) & \gamma_x(0) \end{bmatrix} \begin{bmatrix} b_1 \\ b_2 \\ b_3 \end{bmatrix} = \begin{bmatrix} \gamma_x(-3) \\ \gamma_x(-2) \\ \gamma_x(-1) \end{bmatrix}$$

$$\Rightarrow \begin{bmatrix} \gamma_x(2) & \gamma_x(1) & \gamma_x(0) \\ \gamma_x(1) & \gamma_x(0) & \gamma_x(-1) \\ \gamma_x(0) & \gamma_x(-1) & \gamma_x(-2) \end{bmatrix} \begin{bmatrix} b_1 \\ b_2 \\ b_3 \end{bmatrix} = \begin{bmatrix} \gamma_x(-1) \\ \gamma_x(-2) \\ \gamma_x(-3) \end{bmatrix}$$

Let us represent $\begin{bmatrix} \gamma_x(1) \\ \gamma_x(2) \\ \gamma_x(3) \end{bmatrix}$ as the following:

$$a_1 \begin{bmatrix} \gamma_x(0) \\ \gamma_x(1) \\ \gamma_x(2) \end{bmatrix} + a_2 \begin{bmatrix} \gamma_x(-1) \\ \gamma_x(0) \\ \gamma_x(1) \end{bmatrix} + a_3 \begin{bmatrix} \gamma_x(-2) \\ \gamma_x(-1) \\ \gamma_x(0) \end{bmatrix} \tag{5.12}$$

Similarly we represent $\begin{bmatrix} \gamma_x(-1) \\ \gamma_x(-2) \\ \gamma_x(-3) \end{bmatrix}$ as the following:

$$b_1 \begin{bmatrix} \gamma_x(2) \\ \gamma_x(1) \\ \gamma_x(0) \end{bmatrix} + b_2 \begin{bmatrix} \gamma_x(1) \\ \gamma_x(0) \\ \gamma_x(-1) \end{bmatrix} + b_3 \begin{bmatrix} \gamma_x(0) \\ \gamma_x(-1) \\ \gamma_x(-2) \end{bmatrix} \tag{5.13}$$

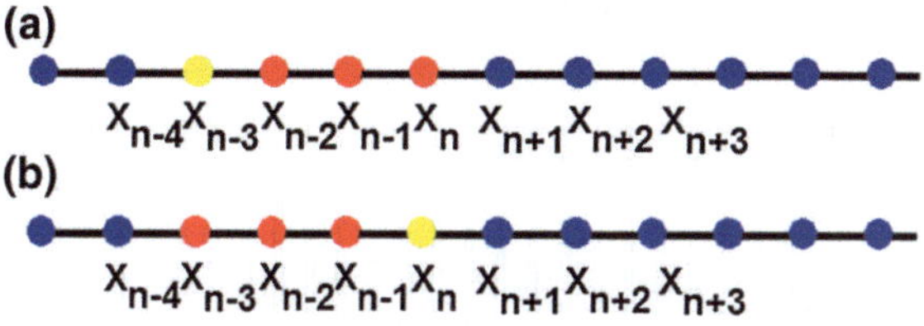

Fig. 5.3 Illustration of the **a** forward prediction, **b** backward prediction (The yellow sample is predicted using the red samples)

From (5.12) and (5.13), we observe $b_m = a_{4-m}$, where 4 is the number of coefficients +1 used. In general for complex samples and with r number of coefficients, we get $b_m = a^*_{r-m}$. The coefficients a_m are known as forward predication coefficients and b_m are known as backward prediction coefficients.

5.2.1.1 Forward and Backward Prediction

In general, we represent the prediction of X_n using one previous sample as $\widehat{X}_n^{f,1} = a_1^{f,1} X_{n-1}$ and the corresponding forward prediction error is given as $e_n^{f,1} = X_n - a_{f,1}^1 X_{n-1}$. Similarly, the prediction of X_n using two previous samples are given as $\widehat{X}_n^{f,2} = a_{f,1}^2 X_{n-1} + a_{f,2}^2 X_{n-2}$. In general, the prediction of X_n using previous M samples and the corresponding forward prediction error is given as the following:

$$\widehat{X}_n^{f,M} = \sum_{k=1}^{k=M} a_{f,k}^M X_{n-k}$$

$$e_n^{f,M} = X_n - \sum_{k=1}^{k=M} a_{f,k}^M X_{n-k}$$

$$= \sum_{k=0}^{k=M} c_{f,k}^M X_{n-k}$$

with $c_{f,k}^M = 1$ for $k = 0$, $c_{f,k}^M = -b_{f,k}^M$ for $k \neq 0$ and $k = 1 \cdots M$.

In the same fashion, we represent the prediction of X_{n-1} using one next sample and it is given as $\widehat{X}_{n-1}^{b,1} = a_1^{b,1} X_n$ and the corresponding backward prediction error is given as the following: $e_n^{b,1} = X_{n-1} - a_1^{b,1} X_n$. In general, the prediction of X_{n-M} using next M samples and the corresponding backward prediction error is given as the following:

$$\widehat{X}_{n-M}^{b,M} = \sum_{k=0}^{k=M-1} a_{b,k}^M X_{n-k}$$

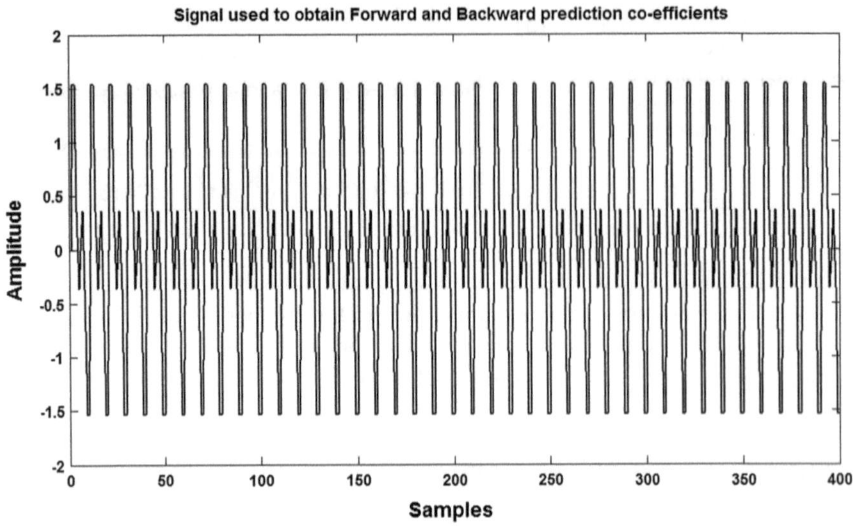

Fig. 5.4 Signal used to compute forward and backward prediction coefficients

$$e_n^{b,M} = X_{n-M} - \sum_{k=1}^{k=M} a_{b,k}^M X_{n-k}$$

$$e_n^{b,M} = \sum_{k=0}^{k=M} c_{b,k}^M X_{n-k}^M$$

with $c_{b,k}^M = 1$ for $k = M$, $c_{b,k}^M = -a_{b,k}^M$ for $k \neq M$ and $k = 1 \cdots M - 1$.

It is noted that $e_n^{f,M}$ is the error involved in estimating X_n using M previous samples and $e_n^{f,M}$ is the error involved in estimating X_{n-M} using the next M samples. They are, respectively, forward prediction error and the backward prediction error (Figs. 5.3, 5.4, 5.5, 5.6, 5.7, 5.8).

5.2.1.2 Methodology Used to Obtain the Coefficients $c_{b,k}^M$ and $c_{f,k}^M$ Using the Coefficients $c_{b,k}^{M-1}$ and $c_{f,k}^{M-1}$

From Sect. 5.2.1.1, we get $c_{f,k}^M = \overset{*}{c}_{b,M-k}^{\;\;M}$. In z-transformation, we get the following:

$$c_f^M(z) = c_b^M(\frac{1}{z})z^{-M}$$

We represent the following:

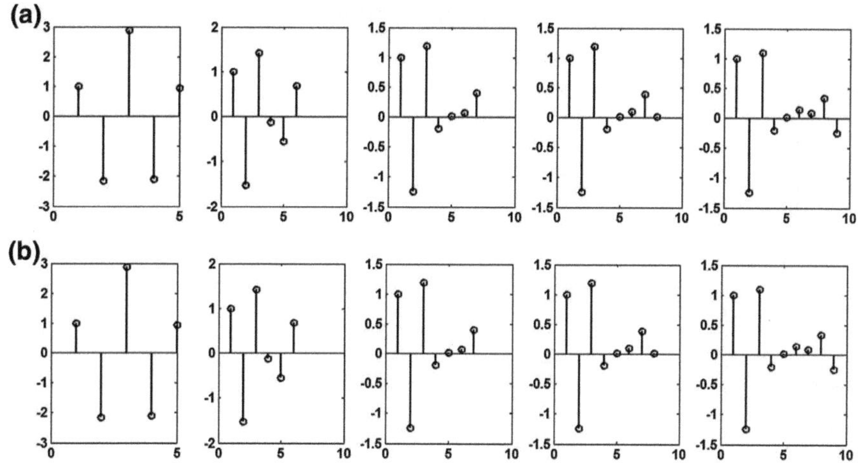

Fig. 5.5 **a** Forward prediction coefficient computed directly by constructing toeplitz matrix. **b** Forward prediction coefficient computed using recursive techniques. Number of co-efficients are increasing from left to right

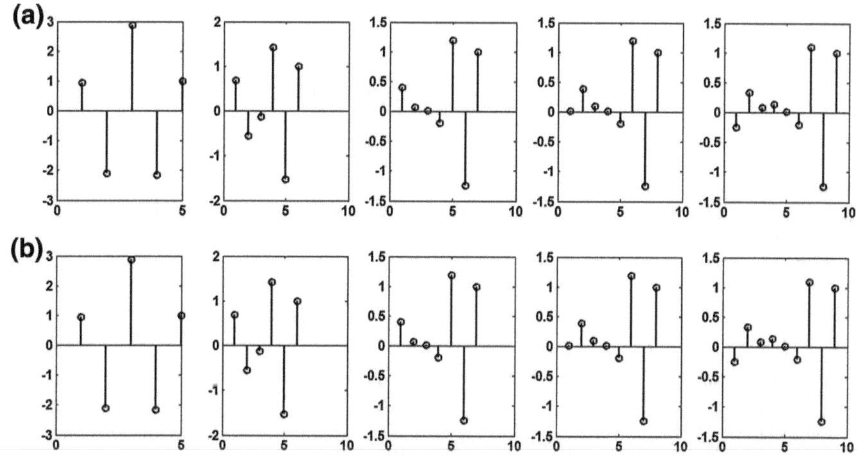

Fig. 5.6 **a** Backward prediction coefficient computed directly by constructing toeplitz matrix. **b** Backward prediction coefficient computed using recursive techniques. Number of co-efficients are increasing from left to right

$$c_{f,k}^{M} = c_{f,k}^{M-1} + K_M c_{b,k-1}^{M-1}$$
$$c_{b,k}^{M} = K_M^* c_{f,k}^{M-1} + c_{b,k-1}^{M-1}$$

where $K_M = c_{f,M}^{M}$ is the reflection coefficient. Taking z-transformation, we get the following:

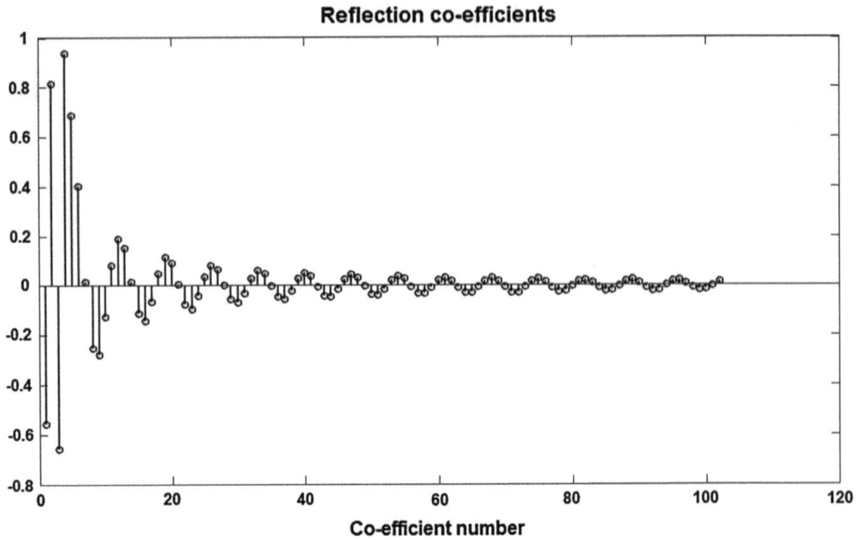

Fig. 5.7 Reflection coefficients obtained for the signal in Fig. 5.4

$$C_f^M(z) = c_f^{M-1}(z) + z^{-1}K_M C_b^{M-1}$$
$$C_b^M = K_M^* C_f^{M-1}(z) + z^{-1}c_b^{M-1}$$

```
%forbackcoef.m
%Computation of Forward, Backward prediction coefficients and the
%reflection
%Generation of signal
n=0:1:1000;
f1=1;
f2=2;
Ts=1/10;
X=sin(2*pi*f1*n*Ts)+sin(2*pi*f2*n*Ts);
figure(1)
plot(X(1:1:400))
title('Signal used to obtain AR coefficients')
R=xcorr(X);
for recur=0:1:101
m1=toeplitz(R(1001:1:1001+recur));
%Computing forward coefficients
v1=[R(1002:1:1002+recur)];
forwardcoef{recur+1}=[1; -1*(inv(m1)*v1')];
backwardcoef{recur+1}=forwardcoef{recur+1}...
(length(forwardcoef{recur+1}):-1:1);
K(recur+1)=forwardcoef{recur+1}...
(length(forwardcoef{recur+1}));
end
F{1}=forwardcoef{1};
B{1}=backwardcoef{1};
for i=4:1:101
%Computing forwardcoef and backwardcoef recursively
F{i}=[F{i-1};0]+K(i)*[0; B{i-1}];
```

```
B{i}=[K(i)*F{i-1};0]+[0; B{i-1}];
end
figure
title('Computation of forward coefficients ...
 using direct and the recursive method')
for i=1:1:5
subplot(2,5,i)
stem(F{i+3})
subplot(2,5,i+5)
stem(forwardcoef{i+3})
end

figure
title('Computation of backward coefficients ...
using direct and the recursive method')
for i=1:1:5
subplot(2,5,i)
stem(B{i+3})
subplot(2,5,i+5)
stem(backwardcoef{i+3})
end
figure
stem(K)
title('Reflection coefficients')

Errorsignal=conv(F{101},X);
v=var(Errorsignal,1);
NOISE=sqrt(v)*rand(1,1000);
RES=[zeros(1,101) X(2)];
for i=1:1:1000
temp1=RES(length(RES):-1:length(RES)-100);
temp2=NOISE(i)-sum(temp1'.*F{101}(2:1:length(F{101})));
RES=[RES temp2];
end
figure
subplot(3,1,1)
plot(NOISE(101:1:200))
title('Input white noise signal with variance 0.0044')
subplot(3,1,2)
plot(X(1:1:100))
title('Actual signal used to obtain coefficients')
subplot(3,1,3)
plot(RES(101:1:200))
title('Output of the AR model')
```

5.3 Adaptive Filter

5.3.1 System Model

Let the Wide Stationary Stationary random process associated with the input to the unknown system is represented as X_n and the corresponding output random process is represented as Y_n. We would like to model the system using FIR filter. Let the filter coefficients be represented as $\mathbf{h} = [h_1 \ h_2 \ \cdots h_{N-1}]^T$. The relationship between Y_n and X_n is as given below.

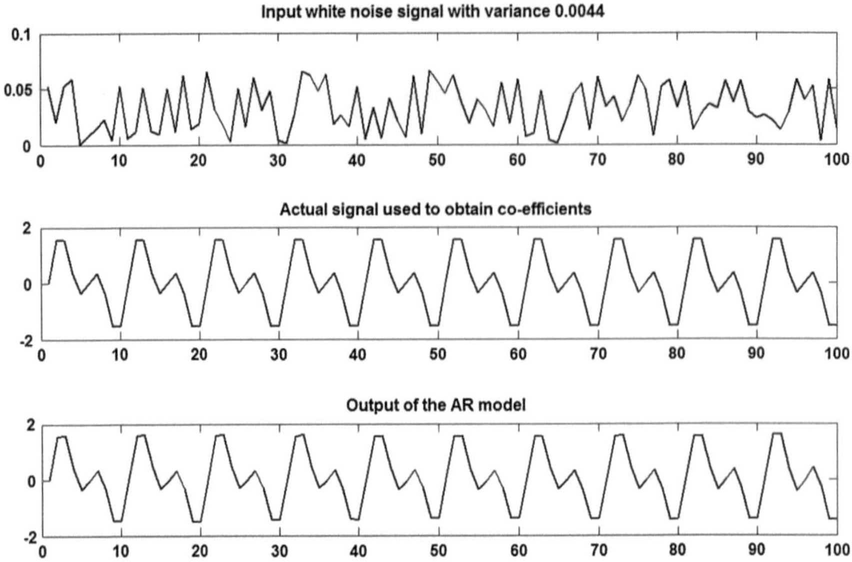

Fig. 5.8 Demonstration of generation of AR random process with white noise as the input

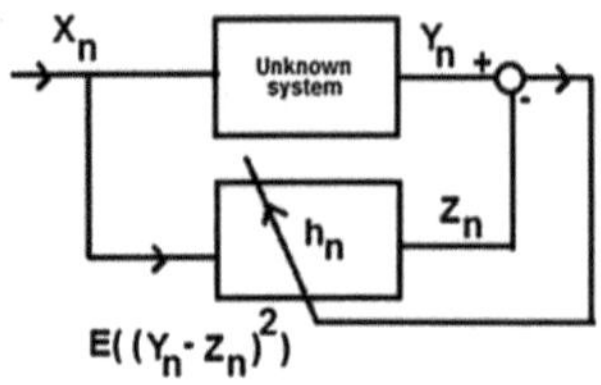

Fig. 5.9 Adaptive filter to model the unknown system

$$Y_n = \sum_{k=0}^{N-1} h_k X_{n-k}$$

We would like to optimize the FIR filter coefficients such that

$$\varepsilon = E((Y_n - \sum_{k=0}^{N-1} h_k X_{n-k})^2) \tag{5.14}$$

is minimized (Fig. 5.9).

Differentiating (5.14), with respect to h_r and equate to zero, we get the following: for $r = 1 \cdots N-1$.

$$E(2(Y_n - \sum_{k=0}^{N-1} h_k X_{n-k})X_{n-r}) = 0 \tag{5.15}$$

$$\gamma_{yx}(r) = \sum_{k=0}^{N-1} h_k \gamma_x(r-k) \tag{5.16}$$

$$\Rightarrow \gamma_{\mathbf{yx}} = \Gamma_x \mathbf{h}_{\mathbf{optimum}} \tag{5.17}$$

$$\Rightarrow \mathbf{h}_{\mathbf{optimum}} = \Gamma_x^{-1} \gamma_{\mathbf{yx}} \tag{5.18}$$

The above is the Wiener–Hopf equation. To solve (5.18), we need the joint pdf $f_{Y_n X n}(\alpha, \beta)$, which is usually not available. Hence iterative technique based on the steepest-descent algorithm is used and the outline of the algorithm is described below:

We would like to get the optimal solution for the variable x such that the function $f(x)$ is minimized. Steepest-descent algorithm starts with randomly chosen initial value and ($x_{current}$). The slope of the function $f(x)$ at $x_{current}$ is computed. From the figure, it is seen that the if the slope is negative, the next best choice (x_{next}) is greater than $x_{initial}$ and is given $x_{next} = x_{current} - \alpha\frac{df(x)}{dx}$, where α is the learning constant. It is noted that if the slope is negative, the x_{next} is greater than $x_{current}$ and vice versa. In case of higher dimension, the iterative equation to obtain the minima point is represented as $\mathbf{x}_{next} = \mathbf{x}_{current} - \nabla \mathbf{f}$, where $\nabla \mathbf{f}$ is the gradient of the function f. Note that x used in this paragraph is the unknown variable/vector to be optimized using steepest-descent algorithm (Fig. 5.10).

The steepest-descent algorithm to optimize $\mathbf{h}$ in (5.14) is obtained by computing the gradient of (5.14) as follows. Let $\mathbf{x} = [x_n \ x_{n-1} \cdots x_{n-N+1}]^T$ and (5.14) is written as the following:

$$\varepsilon = E((Y_n - \sum_{k=0}^{N-1} h_k X_{n-k})^2) = E((Y_n - \mathbf{h}^T\mathbf{x})^2) \tag{5.19}$$

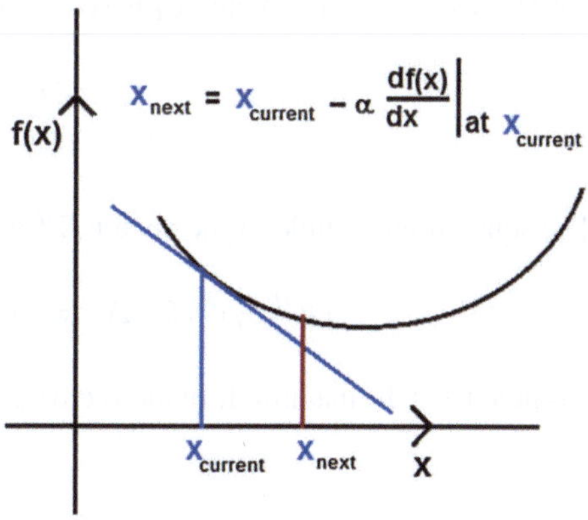

Fig. 5.10 Illustration of the steepest-descent algorithm

Differentiating (5.19) with respect to **h**, we get the following:

$$\nabla\varepsilon = E(-2(Y_n - \mathbf{h}^T\mathbf{x})\mathbf{x})$$
$$\nabla\varepsilon = E(-2\mathbf{e_n}\mathbf{x}) = -2E(\mathbf{e_n}\mathbf{x})$$

Thus the steepest-descent iterative equation is as follows.

$$\mathbf{h}_{t+1} = \mathbf{h}_t - \frac{1}{2}\alpha\ \nabla\varepsilon = \mathbf{h}_t + \alpha E(\mathbf{e_n}\mathbf{x}) \tag{5.20}$$

where t is the tth iteration. The difference between the $\mathbf{h_{t+1}} - \mathbf{h_{optimum}}$ is the error vector (Δ_{t+1}), which denote how the vector attained in the $(t+1)^{th}$ iteration $(\mathbf{h_{t+1}})$ is deviated from the $\mathbf{h_{optimum}}$.

$$\begin{aligned}
&\mathbf{h_{t+1}} - \mathbf{h_{optimum}}\\
&= \mathbf{h}_t + \alpha E(\mathbf{e_n}\mathbf{x}) - \mathbf{h_{optimum}}\\
&= \mathbf{h}_t + \alpha E((Y_n - \mathbf{h_t}^T\mathbf{x})\mathbf{x}) - \mathbf{h_{optimum}}\\
&= \mathbf{h}_t + \alpha E((Y_n - \mathbf{x}^T\mathbf{h_t})\mathbf{x}) - \mathbf{h_{optimum}}\\
&= \mathbf{h}_t + \alpha(\gamma_{\mathbf{yx}} - \Gamma_x\mathbf{h_t}) - \mathbf{h_{optimum}}\\
= (I - \alpha\Gamma_x)\mathbf{h}_t + \alpha\Gamma_x h_{optimum} - \mathbf{h_{optimum}} &= (I - \alpha\Gamma_x)\mathbf{h}_t - (I - \alpha\Gamma_x)h_{optimum}\\
&= (I - \alpha\Gamma_x)(\mathbf{h}_t - h_{optimum})
\end{aligned}$$

Let the deviation in the tth iteration is represented as $\Delta_t = \mathbf{h}_t - h_{optimum}$.

$$\Delta_{t+1} = (I - \alpha\Gamma_x)\Delta_t \tag{5.21}$$
$$\Rightarrow \Delta_{t+1} = (EIE^H - \alpha EDE^H)\Delta_t \tag{5.22}$$
$$\Rightarrow \Delta_{t+1} = E(I - \alpha D)E^H\Delta_t \tag{5.23}$$
$$\Rightarrow E^H\Delta_{t+1} = (I - \alpha D)E^H\Delta_t \tag{5.24}$$

Let $\mathbf{U_t} = E^H\Delta_{t+1}$ and hence $\mathbf{U_t}$ is solved as given below.

$$U_{t+1} = (I - \alpha D)\mathbf{U_t} \tag{5.25}$$
$$U_t = (I - \alpha D)^t U_0 \tag{5.26}$$

The squared magnitude of the vector $E^H\Delta_t$ is computed as

$$(E^H\Delta_t)^H(E^H\Delta_t) = \Delta_t^H EE^H\Delta_t = \Delta_t^H\Delta_t$$

It is noted that the magnitude of the vector $E^H\Delta_t$ is identical with that of the vector Δ_t.

Hence minimizing the kth element of the error vector Δ_t is identical with minimizing the kth element of the vector $\mathbf{U_t}$, i.e., $\mathbf{U_t}(k)$.

$$U_t(k) = (1 - \alpha\lambda_k)^t U_0(k)$$

If $|1 - \alpha\lambda_k| \leq 1$, $Lt_{t\to\infty}(1 - \alpha\lambda_k)^t U_0(k) = 0$. This implies that for the convergence, α is chosen satisfying $0 \leq \alpha \leq \frac{1}{\lambda_k}$ for $k = 0 \cdots N-1$, which in further implies that the α satisfies $0 \leq \alpha \leq \frac{1}{\lambda_{max}}$.

It is observed from (5.20) that the update equation requires $E(e_n\mathbf{x})$, which in further requires, joint density function of the random variable X. This is rarely available in the real time applications. Hence the estimate of the $E(e_n\mathbf{x})$ is used. It is computed as follows. Considering the L outcomes of the along the process such that $e_n, e_{n-1}, \cdots e_{n-L+1}$ and the corresponding random vectors are represented as $\mathbf{x_n}, \mathbf{x_{n-1}}, \cdots \mathbf{x_{n-L+1}}$ are used to estimate $E(e_n\mathbf{x})$ as follows.

$$\widehat{E}(e_n\mathbf{x}) = \sum_{l=0}^{l=L-1} e_{n-l}\mathbf{x_{n-l}}$$

For the special case when $l = 0$, the estimate is computed as $e_n\mathbf{x_n} = e_n\mathbf{x}$. The iterative equation (5.20) is modified by replacing the actual Expectation with the estimation to obtain the least mean square (LMS) algorithm. Thus the iterative equation for the LMS algorithm is given as the following:

$$\mathbf{h}_{t+1} = \mathbf{h}_t + \alpha\mathbf{e_n}\mathbf{x} \tag{5.27}$$

It is noted that the Expectation of the estimate $\mathbf{e_n}\mathbf{x}$ gives the actual gradient used in the steepest-descent algorithm. Hence the estimate is unbiased with the value used in the steepest-descent algorithm. We also understand intuitively that as an average LMS algorithm behaves like the steepest-descent algorithm. Taking Expectation on both sides of (5.27), we get the following:

$$E(\mathbf{h}_{t+1}) = E(\mathbf{h}_t) + \alpha E(\mathbf{e_n}\mathbf{x})$$
$$= E(\mathbf{h}_t) + \alpha E((\mathbf{Y_n} - \mathbf{h_t^T}\mathbf{x})\mathbf{x})$$
$$= E(\mathbf{h}_t) + \alpha\gamma_{yx} - \alpha\Gamma_x E(\mathbf{h}_t)$$

The difference between $\mathbf{h_t}$ and $\mathbf{h_{optimum}}$ is represented as Δ_t. i.e,

$$\Delta_{t+1} = \mathbf{h_{t+1}} - \mathbf{h_{optimum}}$$
$$\Rightarrow E(\Delta_{t+1}) = E(\mathbf{h_{t+1}}) - E(\mathbf{h_{optimum}}) = E(\mathbf{h_{t+1}}) - \mathbf{h_{optimum}}$$

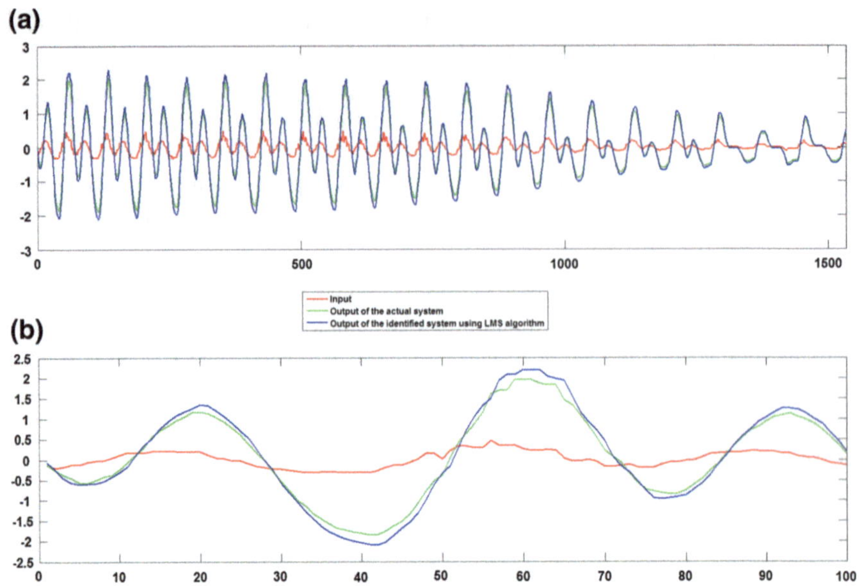

Fig. 5.11 Demonstration of adaptive filter to identify the unknown system

From (5.23), we get the following:

$$
\begin{aligned}
E(\Delta_{t+1}) &= E(\mathbf{h}_t) + \alpha\gamma_{yx} - \alpha\Gamma_x E(\mathbf{h}_t) - \mathbf{h_{optimum}} \\
E(\Delta_{t+1}) &= E(\Delta_t) + \alpha\Gamma_x \mathbf{h_{optimum}} - \alpha\Gamma_x E(\mathbf{h}_t) \\
E(\Delta_{t+1}) &= E(\Delta_t) + \alpha\Gamma_x(\mathbf{h_{optimum}} - E(\mathbf{h}_t)) \\
E(\Delta_{t+1}) &= E(\Delta_t) - \alpha\Gamma_x E(\Delta_t) \\
E(\Delta_{t+1}) &= (I - \alpha\Gamma_x)E(\Delta_t) \\
E(\Delta_{t+1}) &= (\mathbf{v}\mathbf{v^H} - \alpha\mathbf{v}\Lambda\mathbf{v^H})E(\Delta_t) \\
E(\Delta_{t+1}) &= (\mathbf{v}(I - \alpha\Lambda)\mathbf{v^H})E(\Delta_t) \\
\mathbf{v^H}E(\Delta_{t+1}) &= (I - \alpha\Lambda)V^H E(\Delta_t)
\end{aligned}
$$

Let $\mathbf{u^H} = \mathbf{v^H}E(\Delta_{t+1})$, we get the following:

$$\mathbf{u^H_{t+1}} = (I - \alpha\Lambda)\mathbf{u^H_t} \tag{5.28}$$

$$\Rightarrow \mathbf{u^H_{t+1}} = (I - \alpha\Lambda)^t\mathbf{u^H_0} \tag{5.29}$$

The kth element of the vector $\mathbf{u^H}_{t+1}$ is represented as $\mathbf{u}^H_{t+1,k} = (1 - \alpha\lambda_k)^t\mathbf{u^H}_{0,k}$, where $\mathbf{u^H}_{0,k}$ is the kth element of the vector $\mathbf{u^H_0}$. The value of $\mathbf{u}^H_{t+1,k}$ tends to zero as t tends to ∞ for every k, provided $|\alpha\lambda_k| \leq 1$. This implies $|\alpha| \leq \left|\frac{1}{\lambda_k}\right|$ for all k. It is also important to note that the random vector $\mathbf{h}$ and $\mathbf{x}$ are assumed as the independent random vector. Figure 5.11 demonstrates the adaptive filter used to identify the unknown system.

```
%adapsysiden.m
%Adaptive filter
%Identifying the unknown system
%Generation of signal
load speechdata
X=speechdata(2000:1:4999)
Ts=1/8000;
%Let the sysem be represented as h
h=rand(1,11);
%Frequency response of the system under consideration
%Noisy observation is given as the following:
Y=conv(X,h);
%System model using LMS algorithm
X1=reshape(X,100,30);
C=cov(X1);
[E,D]=eig(C);
lambda=diag(D);
lambdamax=lambda(30);
%Fix the learning constant
alpha=1/lambdamax;
%Actual LMS algorithm starts
hlms=rand(1,11);
E=[];
for n=11:1:length(X)
out=X(n-10)*hlms(11)+X(n-9)*hlms(10)+X(n-8)*hlms(9)+ ...
X(n-7)*hlms(8)+ X(n-6)*hlms(7)+ X(n-5)*hlms(6)+...
X(n-4)*hlms(5)+X(n-3)*hlms(4)+X(n-2)*hlms(3)+...
X(n-1)*hlms(2)+X(n)*hlms(1);
error= Y(n)-out;
E=[E error];
hlms=hlms+alpha*error*X(n);
end
Z=conv(X,hlms);
figure(1)
subplot(2,1,1)
plot(X,'r')
hold on
plot(Y,'g')
hold on
plot(Z,'b')
subplot(2,1,2)
plot(X(1:1:100),'r')
hold on
plot(Y(1:1:100),'g')
hold on
plot(Z(1:1:100),'b')
```

5.3.2 Filtering Noise

Consider the random process $Y_n = X_n + W_n$, where X_n and W_n are independent. We would like to estimate W_n by filtering the Y_{n-l} using the linear filter h_k. The filter coefficients are optimized using steepest-descent algorithm by minimizing the objective function as described below:

$$J = E((X_n - \widehat{X}_n)^2) \tag{5.30}$$

$$E((X_n - \sum_{k=0}^{k=N-1} h_k Y_{n-k-l})^2) \tag{5.31}$$

Minimizing J is equivalent to minimizing $E((Y_n - \sum_{k=0}^{k=N-1} h_k Y_{n-k-l})^2)$ as described below:

$$E((Y_n - \sum_{k=0}^{k=N-1} h_k Y_{n-k-l})^2)$$
$$= E((X_n + W_n - \sum_{k=0}^{k=N-1} h_k Y_{n-k-l})^2)$$
$$= E((X_n - \sum_{k=0}^{k=N-1} h_k Y_{n-k-l})^2) + E((W_n)^2) - 2E((W_n - \sum_{k=0}^{k=N-1} h_k Y_{n-k-l})W_n)$$

The second term is constant Consider the third term.

$$E((W_n - \sum_{k=0}^{k=N-1} h_k Y_{n-k-l})W_n) \tag{5.32}$$

$$= E(W_n W_n) - E(\sum_{k=0}^{k=N-1} h_k (X_{n-k-l} + W_{n-k-l})W_n) \tag{5.33}$$

$$= E(W_n^2) - E(\sum_{k=0}^{k=N-1} h_k (X_{n-k-l} W_n)) + E(W_{n-k-l} W_n) \tag{5.34}$$

As X_n and W_n are independent with $E(W_n) = 0$, $E(\sum_{k=0}^{k=N-1} h_k (X_{n-k-l} W_n)) = 0$. Also considering $E(W_{n-k-l} W_n) = 0$, we get the third term as constant as $E(W_n^2)$. Thus the optimal $h(k)$ that minimizes $E((X_n - \sum_{k=0}^{k=N-1} h_k Y_{n-k-l})^2)$ is equivalent to minimizing $E((Y_n - \sum_{k=0}^{k=N-1} h_k Y_{n-k-l})^2)$. Thus the adaptive filter to filter the noise is constructed as given in the Fig. 5.12 and the corresponding illustration is given in Fig. 5.13.

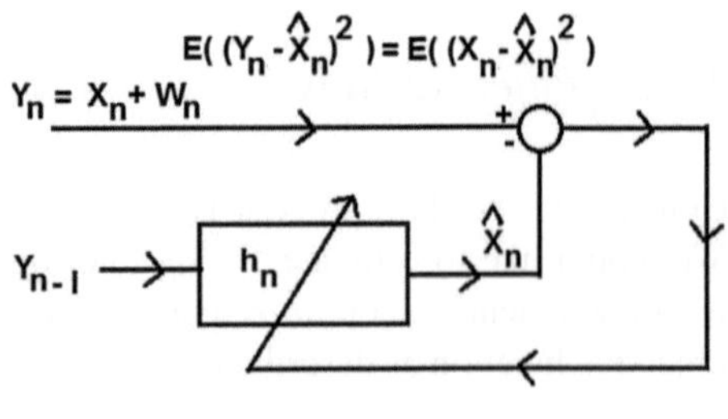

Fig. 5.12 Adaptive filter constructed to filter the additive noise

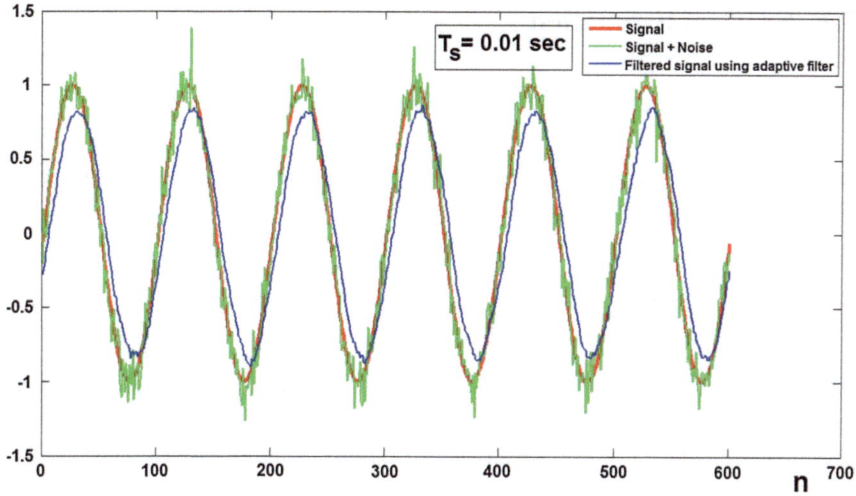

Fig. 5.13 Demonstration of noise removal using the adaptive filter

```
%adapfiltnoise.m
f=1;
Ts=1/100;
n=0:1:1999;
X=sin(2*pi*f*n*Ts);
Y=X+0.1*randn(1,length(X));
Z=0.1*[randn(1,10) Y ];
hadapt=rand(1,11);
for n=12:1:length(Y)
output=hadapt(11)*Z(n-10)+hadapt(10)*Z(n-9)+hadapt(9)*Z(n-8)+...
hadapt(8)*Z(n-7)+hadapt(7)*Z(n-6)+hadapt(6)*Z(n-5)+hadapt(5)*Z(n-4)+
hadapt(4)*Z(n-3)+hadapt(3)*Z(n-2)++hadapt(2)*Z(n-1)+hadapt(1)*Z(n);
error=output-Y(n);
hadapt=hadapt-0.1*error*Z(n);
end
OUTPUT=[];
for n=12:1:length(Y)
output=hadapt(11)*Z(n-10)+hadapt(10)*Z(n-9)+hadapt(9)*Z(n-8)+...
hadapt(8)*Z(n-7)+hadapt(7)*Z(n-6)+hadapt(6)*Z(n-5)+hadapt(5)*Z(n-4)+
hadapt(4)*Z(n-3)+hadapt(3)*Z(n-2)++hadapt(2)*Z(n-1)+hadapt(1)*Z(n);
OUTPUT=[OUTPUT output];
end
figure(2)
plot(X(900:1:1500),'r')
hold on
plot(Y(900:1:1500),'g')
plot(OUTPUT(900:1:1500),'b')
```

5.4 Spectral Estimation

5.4.1 Eigen Decomposition Method for Spectral Estimation

Let the outcome of the discrete random process with linear combinations of P complex single tone signals is represented as the following:

$$x_n = \sum_{m=1}^{m=P} A_m e(j2\pi f_m n - \phi_m) \tag{5.35}$$

where A_m is the coefficient of the complex single tone signal with frequency f_m and ϕ_m is uniformly distributed between 0 to 2π. The auto-correlation of the discrete random process x_n represented as $R_x(k)$ is computed as follows.

$$R_x(k) = E(x_{n+k}x_n^*)$$
$$\Rightarrow R_x(k) = E(\sum_{r=1}^{r=P} A_r e^{j(2\pi f_r(n+k)-\phi_r)} \sum_{s=1}^{s=P} A_s^* e^{-j(2\pi f_s(n)-\phi_s)})$$
$$\Rightarrow R_x(k) = E(\sum_{r=1}^{r=P}\sum_{s=1}^{s=P} A_r A_s^* e^{j(2\pi f_r(n+k)-\phi_r)} e^{-j(2\pi f_s(n)-\phi_s)})$$
$$\Rightarrow R_x(k) = E(\sum_{r=1}^{r=P}\sum_{s=1}^{s=P} A_r A_s^* e^{j(2\pi f_r(n+k)-\phi_r)} e^{-j(2\pi f_s(n)-\phi_s)})$$

Consider the term with $r \neq s$, $E(A_r A_s^* e^{j(2\pi f_r(n+k)-\phi_r)} e^{-j(2\pi f_s(n)-\phi_s)})$ is computed as follows.

$$E(A_r A_s^* e^{j(2\pi f_r(n+k)-\phi_r)} e^{-j(2\pi f_s(n)-\phi_s)})$$
$$\Rightarrow E(A_r A_s^* e^{j(2\pi f_r(n+k)-\phi_r-2\pi f_s(n)+\phi_s)})$$

Let $\phi = \phi_s - \phi_r$ is distributed as shown in the Fig. 5.14. This implies the following:

$$E(A_r A_s^* e^{j(2\pi f_r(n+k)+\phi-2\pi f_s(n))})$$
$$= E(A_r A_s^* e^{j(2\pi(f_r-f_s)n+2\pi f_r k+\phi)})$$

This is solved by obtaining the expectation by fixing ϕ and followed by computing expectation over ϕ (for $r \neq s$) as follows.

$$A_r A_s \int_{-2\pi}^{2\pi} e^{j(2\pi(f_r-f_s)n+2\pi f_r k+\phi)} d\phi = 0$$

Considering the term with $r = s$, we get the following:

$$E(A_r A_r^* e^{j(2\pi f_r(n+k)-\phi_r)} e^{-j(2\pi f_r(n)-\phi_r)})$$
$$\Rightarrow |A_r|^2 e^{j(2\pi f_r k)}$$

Thus $R_x(k)$ computed for the discrete random process X_n in (5.35) is obtained as

$$R_x(k) = \sum_{m=1}^{m=P} |A_m|^2 e^{j(2\pi f_m k)} \tag{5.36}$$

It is also noted that discrete random process x_n in (5.35) is the W.S.S. Random process. Given the typical values for $f_m's$, $R_x(k)'s$, and P, we obtain $|A_m|^2$ using (5.36).

The random process x_n which is the linear combinations of complex single tone frequency signals with different phase and amplitude can be generated as

$$x_n = -\sum_{m=1}^{m=P} v_m x_{n-m} \tag{5.37}$$

Let us define the random process $y_n = x_n + w_n$, where w_n is the additive white noise with variance σ_w^2. The signal space and the noise space are obtained as follows:

$$\sum_{m=0}^{m=P} v_m x_{n-m} = 0 \tag{5.38}$$

$$\Rightarrow \sum_{m=0}^{m=P} v_m(y_{n-m} - w_{n-m})$$

$$\Rightarrow \sum_{m=0}^{m=P} v_m y_{n-m} = \sum_{m=0}^{m=P} v_m w_{n-m}$$

$$\Rightarrow [y_n \; y_{n-1} \; \cdots y_{n-P}][v_0 \; v_1 \; \cdots v_P]^T = [w_n \; w_{n-1} \; \cdots w_{n-P}][v_0 \; v_1 \; \cdots v_P]^T$$

Let $\mathbf{y} = [y_n \; y_{n-1} \; \cdots y_{n-P}]^T$, $\mathbf{w} = [w_n \; w_{n-1} \; \cdots w_{n-P}]^T$ and $\mathbf{v} = [v_0 \; v_1 \; \cdots v_P]^T$ and $\mathbf{x} = [x_n \; x_{n-1} \; \cdots x_{n-P}]^T$, we get the following:

$$\mathbf{y}^T\mathbf{v} = \mathbf{w}T\mathbf{v}$$
$$\mathbf{y}\mathbf{y}^T\mathbf{v} = \mathbf{y}\mathbf{w}T\mathbf{v}$$

Taking Expectations on both sides (Note that the mean of the random vectors $\mathbf{y}$ and $\mathbf{w}$ are identically zero)

$$C_y \mathbf{v} = C_{yw} \mathbf{v}$$
$$C_y \mathbf{v} = \sigma_w^2 I \mathbf{v}$$
$$C_y \mathbf{v} = \sigma_w^2 \mathbf{v}$$

Thus the vector $\mathbf{v}$ (with first element normalized to 1) is the Eigen vector of the matrix C_y corresponding to the Eigen value σ_w^2. The random vector $\mathbf{y} = \mathbf{x} + \mathbf{w}$. This implies the following:

$$C_y = C_x + C_w \tag{5.39}$$
$$C_x + \sigma_w^2 I \tag{5.40}$$

From (5.35), we represent $\mathbf{x} = \sum_{i=1}^{i=P} A_i e^{j2\pi f_i n - \phi_i} \mathbf{s_i}$, where

$$\mathbf{s_i} = [1 \; e^{-j2\pi f_i} \; e^{-j4\pi f_i} \; \cdots \; e^{-j2P\pi f_i}]^T \tag{5.41}$$

Thus we understand that the rank of the symmetric matrix C_x is P (note that the size of the matrix is $(P+1) \times (P+1)$) and hence we get P nonzero Eigenvalues $(\lambda_1, \lambda_2, \cdots \lambda_P)$ and 1 zero Eigenvalue. Thus the Eigen values of the matrix C_y are given as $\lambda_1 + \sigma_w^2, \; \lambda_2 + \sigma_w^2, \; \cdots \; \lambda_P + \sigma_w^2, \; \sigma_w^2$. Let the corresponding Eigenvectors are represented as $\mathbf{e_1}, \mathbf{e_2}, \cdots \mathbf{e_{P+1}}$. From (5.39) and (5.40), we understand that the elements of the Eigenvector of the matrix C_y corresponding to the Eigenvalue σ_w^2 satisfy (5.40). The space spanned by the Eigenvectors $\mathbf{e_1}, \mathbf{e_2}, \cdots \mathbf{e_P}$ is the signal space and the space spanned by the Eigenvector $\mathbf{e_{P+1}}$ is the noise space. Note that the Eigenvector $\mathbf{e_{P+1}}$ satisfies $C_x \mathbf{e_{P+1}} = 0$. It can also be interpreted as the column space of the matrix C_x forms the signal space and the null space of the matrix C_x forms the noise space. It is also understood, the Eigenvector $\mathbf{e_{P+1}}$ corresponding to the eignvalue 0 of the matrix C_x is identical with that of the Eigenvector of the matrix C_y corresponding to the Eigenvalue σ_w^2 and hence $\mathbf{e_{P+1}} = \mathbf{v}$. Thus the Eigenvector $\mathbf{v}$, which is in the noise space is orthogonal to the all the vectors in the signal space. Hence $\mathbf{s_i}^H \mathbf{v} = 0$ for all i. i.e,

$$\sum_{k=0}^{k=P} e^{-j2\pi k f_i} v_k = 0 \tag{5.42}$$

where v_k is the $(k+1)$th element of the vector $\mathbf{v}$ with $v_0 = 1$.

5.4.2 *Pisarenko Harmonic Decomposition*

In (5.35), x_n is viewed as the prediction value at the nth sample instant using the linear combinations of previous P samples. The difference between the actual value and the predicted value is represented as $e_n = x_n + \sum_{m=1}^{m=P} v_m x_{n-m}$. Assuming the

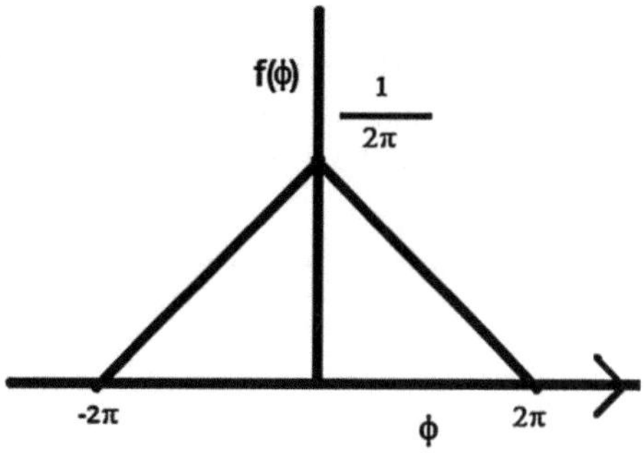

Fig. 5.14 Probability density function of the random variable ϕ

e_n is the input to the system and x_n as the corresponding output of the system (AR system), we get the recursive system function as given below (Fig. 5.14).

$$\frac{X(z)}{E(z)} = \frac{1}{1 + \sum_{m=1}^{m=P} v_m z^{-m}}$$

From (5.37) and (5.40), we understand that the elements of the Eigenvector with leading element 1 of the matrix C_y corresponding to the Eigenvalue σ_w^2 satisfies the polynomial $1 + \sum_{m=1}^{m=P} v_m z^{-m} = 0$. Thus the method of estimating the frequencies are summarized below:

1. Assuming that the given signal is the linear combination of p number of complex single tone signals with additive noise.
2. Estimate the covariance matrix C_y of size $(p+1) \times (p+1)$.
3. Obtain the lowest Eigenvalue and the corresponding vector, with the first element made it as 1. Let it be $\mathbf{v}$.
4. Formulate the polynomial $1 + \sum_{m=1}^{m=P} v_m z^{-m}$ and the P roots of this polynomial are treated as the estimate of P single tone frequencies present in the given noisy signal.

5.4.3 MUSIC (Multiple Signal Classification) Method

Suppose that the covariance matrix is constructed with size $m \times m$ with $m > P$, the dimension of the signal space is P. The Eigen vectors corresponding to the largest P Eigenvalues correspond to the signal space. The remaining Eigen vectors corresponding to $m - P$ Eigenvalues correspond to the noise space. Let the Eigen vectors corresponding to the lowest $m - P$ Eigenvalues are represented as $\mathbf{v_{P+1}} \mathbf{v_{P+2}} \cdots \mathbf{v_m}$. The first element of the individual Eigen vectors are made 1. The Eigen vectors $\mathbf{v_k}$ with $k = P + 1 \cdots m$ satisfy the polynomial $1 + \sum_{r=1}^{r=m-1} v_{k,r} e^{-j2\pi \frac{f}{fs} r} = 0$, where $v_{k,r}$ is the rth element of the kth Eigen vector. From (4) in Pisarenko Harmonic Decomposition technique, the roots of the polynomial $1 + \sum_{r=1}^{r=m-1} v_{k,r} e^{-j2\pi \frac{f}{fs} r} = 0$

are the estimate of the P single tone frequencies present in the given noisy signal, if $m = P + 1$. Hence the estimate of the P single tone frequencies present in the given noisy signal are also obtained by identifying the values for f (frequencies) corresponding to the first P largest magnitude of the polynomial given below:

$$\frac{1}{\sum_{k=P+1}^{m} \left|(1 + \sum_{r=1}^{r=m-1} v_{k,r} e^{-j2\pi \frac{f}{f_s} r})\right|^2}. \tag{5.43}$$

5.4.4 ESPRIT (Estimation of Signal Parameters Via Rotational Invariance Technique)

Consider the following with $P = 2$.

$$\mathbf{x} = [x_n \; x_{n-1} \; x_{n-2}]^T$$
$$\mathbf{y} = \mathbf{x} + \mathbf{w}$$
$$\mathbf{z} = [x_{n+1} \; x_n \; x_{n-1}]^T + [w_{n+1} \; w_n \; w_{n-1}]^T$$

Using $\mathbf{x} = \sum_{i=1}^{i=2} A_i e^{(j2\pi f_i n - \phi_i)} \mathbf{s_i}$, we get the following:

$$\mathbf{x} = [\mathbf{s_1} \; \mathbf{s_2}] \begin{bmatrix} A_1 e^{-j\phi_1} e^{j2\pi f_1 n} \\ A_2 e^{-j\phi_1} e^{j2\pi f_2 n} \end{bmatrix} = \mathbf{Sp}$$

where $\mathbf{s_1}$ and $\mathbf{s_2}$ are given by (5.41). Similarly, $[x_{n+1} \; x_n \; x_{n-1}]^T$ is computed as follows:

$$[x_{n+1} \; x_n \; x_{n-1}]^T = \mathbf{S} \begin{bmatrix} e^{j2\pi f_1} & 0 \\ 0 & e^{j2\pi f_2} \end{bmatrix} \mathbf{p}$$
$$S\phi\mathbf{p}$$

Let $\mathbf{w_1} = [w_{n+1} \; w_n \; w_{n-1}]^T$ Rewriting the expression as the following:

$$\mathbf{y} = \mathbf{Sp} + \mathbf{w}$$
$$\mathbf{z} = \mathbf{S}\phi\mathbf{p} + \mathbf{w_1}$$
$$\Rightarrow \mathbf{C_y} = E(\mathbf{y}\mathbf{y}^H) = E((\mathbf{Sp} + \mathbf{w})(\mathbf{Sp} + \mathbf{w})^H)$$
$$= \mathbf{S}E(PP^H)\mathbf{S}^H + E(\mathbf{w}(\mathbf{w})^H)$$
$$= \mathbf{S}\Gamma\mathbf{S}^H + \sigma_w^2 I$$

We also compute $\mathbf{C_{yz}}$ as follows:

$$\begin{aligned}&E((\mathbf{Sp}+\mathbf{w})(\mathbf{S}\phi\mathbf{p}+\mathbf{w_1})^h)\\&=\mathbf{S}E(PP^H)(\phi)^H\mathbf{S}^H+E((\mathbf{w})(\mathbf{w_1})^H)\\&=\mathbf{S}\Gamma\phi^H\mathbf{S}^H+\Gamma_w\end{aligned}$$

Let $A = C_y - \sigma_w^2 I = \mathbf{S}\Gamma\mathbf{S}^H$ and $B = C_{yx} - \Gamma_w = \mathbf{S}\Gamma\phi^H\mathbf{S}^H$. It is noted that $\mathbf{S}\Gamma\mathbf{S}^H$ is the covariance matrix computed for the vector $\mathbf{x}$ and hence the rank of the matrix $\mathbf{S}\Gamma\mathbf{S}^H$ is 2. It is also noted that the rank of the matrix $\mathbf{S}\Gamma\phi^H\mathbf{S}^H$ is 2.

Consider the following:

$$\begin{aligned}&A-\lambda B\\&=\mathbf{S}\Gamma\mathbf{S}^H-\lambda\mathbf{S}\Gamma\phi^H\mathbf{S}^H\\&=\mathbf{S}\Gamma(I-\lambda\phi^H)\mathbf{S}^H\end{aligned}$$

The set of vectors $\mathbf{v_i}$ satisfying $(A-\lambda B)\mathbf{v_i} = 0$ are the generalized Eigen vectors. The size of the matrix $(A-\lambda B)$ is 3×3 and it's rank is 2. As the rank of the matrix A is 2, there exist the generalized Eigen vector $\mathbf{v_1}$ corresponding to the Eigenvalue $\lambda = 0$. To obtain the generalized Eigen vector, we need to choose value for λ that makes the rank of the matrix $(A-\lambda B)$ 2. If $\lambda = e^{j2\pi f_1}$, we represent the matrix $\mathbf{S}\Gamma(I-\lambda\phi^H)\mathbf{S}^H$ as follows.

$$\mathbf{S}\Gamma\begin{bmatrix}0 & 0\\ 0 & 1-e^{j2\pi f_1}e^{-j2\pi f_2}\end{bmatrix}\mathbf{S}^H$$

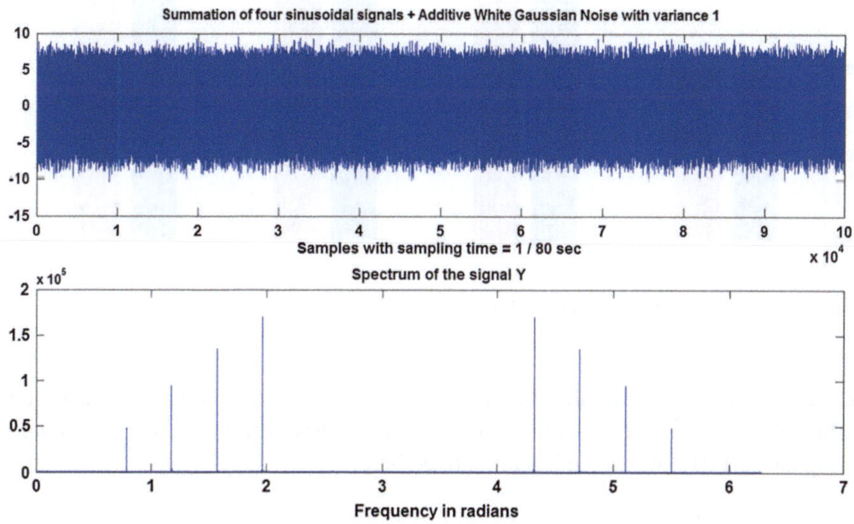

Fig. 5.15 Generated signal to demonstrate eigen decomposition method for spectral estimation

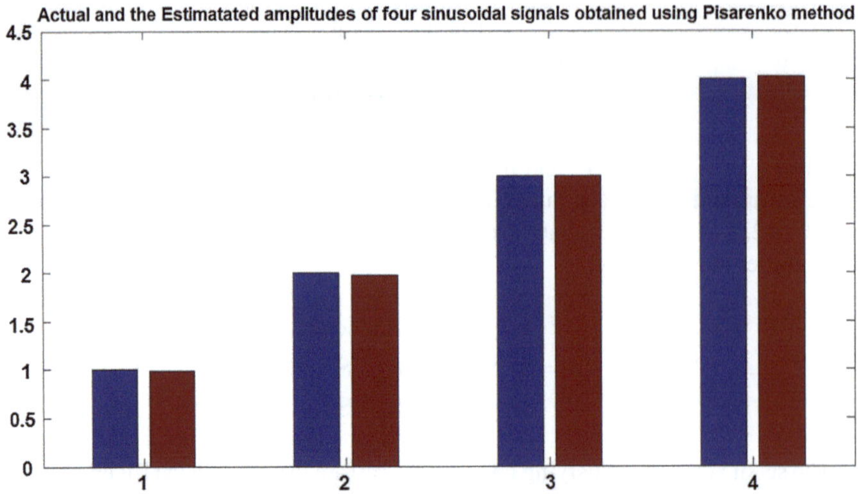

Fig. 5.16 Illustration of Pisarenko method for spectral estimation

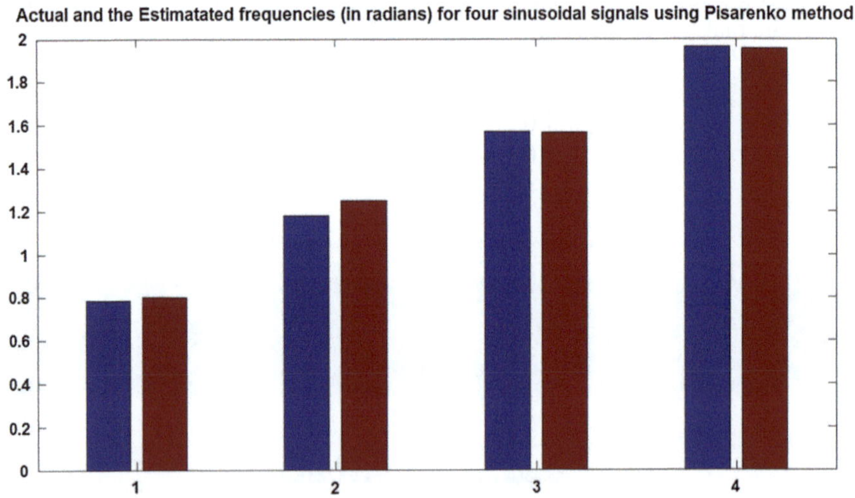

Fig. 5.17 Illustration of Pisarenko method for spectral estimation (continued)

Thus there exists the generalized Eigen vectors $\mathbf{v}_2$ and $\mathbf{v}_3$ corresponding to the Eigenvalues, $\lambda = e^{j2\pi f_1}$ and $\lambda = e^{j2\pi f_2}$. Thus the procedure to estimate the frequencies using the ESPRIT technique is summarized below.

- Estimate the matrix A and B.
- Compute $(A - \lambda B)$.
- Compute the generalized Eigenvalues (with magnitude 1) of the matrix $(A - \lambda B)$, i.e., $(A - \lambda B)\mathbf{v} = 0$.

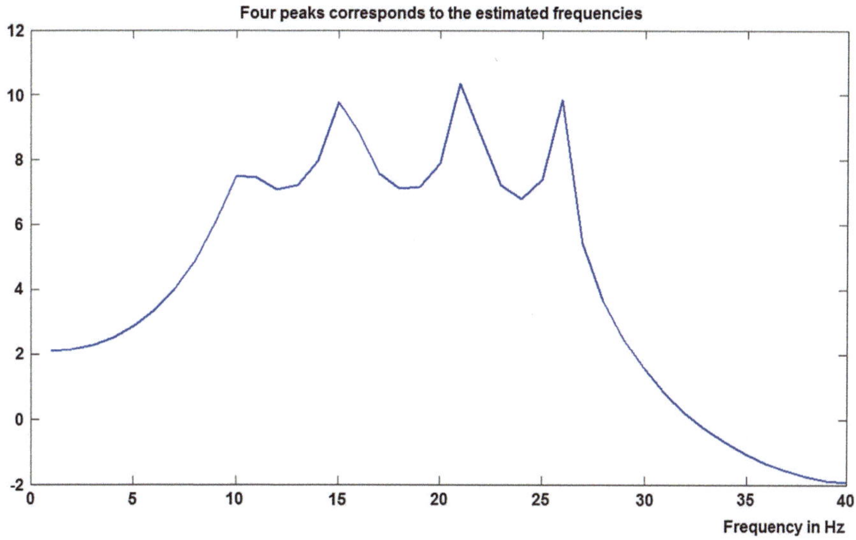

Fig. 5.18 Illustration of MUSIC algorithm

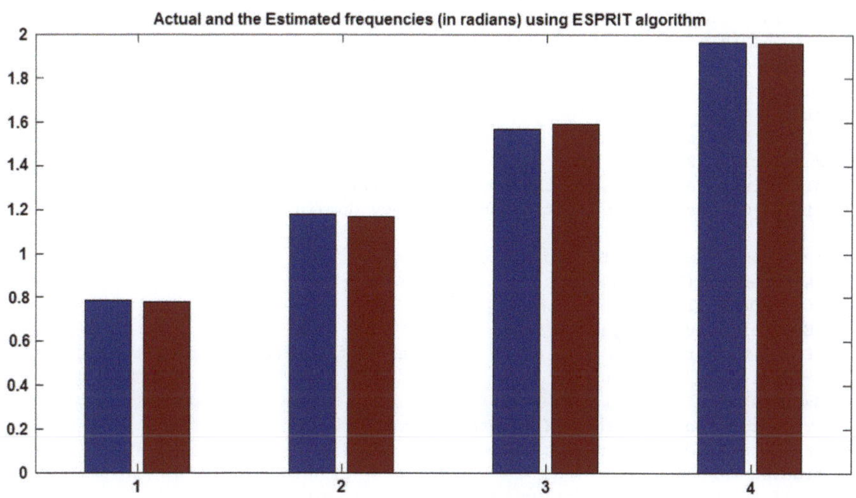

Fig. 5.19 Illustration of ESPRIT algorithm

- Thus the obtained Eigenvalues corresponds to the estimated frequencies. Figures 5.15, 5.16, 5.17, 5.18, 5.19 demonstrates the ESPRIT algorithm.

```
%pisarenko-music-esprit.m
%Demonstration of Pisarenko Harmonic Decomposition method,
%MUSIC  and ESPRIT algorithm
%Generation of linear combinations of sinusoidal signals
```

```
%Y=X+W model
close all
%Input frequency in Hz
f1=10;
f2=15;
f3=20;
f4=25;
F=[f1 f2 f3 f4];
%Amplitude of the input signal
A1=1;
A2=2;
A3=3;
A4=4;
A=[A1 A2 A3 A4];
n=0:1:99998;
fs=80;
ts=1/fs;
%Input frequency in radians
W=(F/fs)*2*pi
%Input
phi1=rand*2*pi;
phi2=rand*2*pi;
phi3=rand*2*pi;
phi4=rand*2*pi;
Y=A1*sin(2*pi*f1*n*ts-phi1)+A2*sin(2*pi*f2*n*ts-phi2)+...
A3*sin(2*pi*f3*n*ts-phi3)+...
A4*sin(2*pi*f4*n*ts-phi4)+sqrt(1)*randn(1,length(n));
figure
subplot(2,1,1)
plot(Y)
title('Summation of three sinusoidal signals with additive gaussian noise')
subplot(2,1,2)
plot(linspace(0,2*pi,length(Y)),abs(fft(Y)))
title('Corresponding spectrum in frequency domain')

%Pisarenko Harmonic Decomposition Method
DATA=reshape(Y,9,11111);
C=cov(DATA',1);
[E,D]=eig(cov(DATA',1));
%Eigen vector in the null space of columns of the covariance matrix C
V=E(:,1);
Z=roots(V);
freqestimation=angle(Z(1:2:length(Z)));
freqestimation=mod(freqestimation,pi)
figure
bar([sort(unwrap(W));sort(unwrap(freqestimation))']')
title('Actual and the Estimatated frequencies using ...
Pisarenko Harmonic Decomposition Method')
MATRIX=[];
for i=1:1:4
MATRIX=[MATRIX; cos(i*W(1)) cos(i*W(2)) cos(i*W(3)) cos(i*W(4))];
end
temp=C(1,:);
GAMMAYY=temp(2:1:5);
powestimation=pinv(MATRIX)*GAMMAYY'
ampestimation=sqrt(abs(2*powestimation));
figure(2)
bar([sort(A);sort(ampestimation)']')
title('Actual and the Estimatated amplitudes using ...
```

```
Pisarenko Harmonic Decomposition Method')

%Multiple Signal Classification (MUSIC) algorithm
Y1=Y(1:1:90000);
DATA=reshape(Y1,9,10000);
C=cov(DATA',1);
[E,D]=eig(cov(DATA',1));
%Eigen vector in the null space of columns of the covariance matrix C
FINAL=0;
for k1=1:1:1
temp1=E(:,k1);
TEMP=[];
for f=0:1:39
s=0;
for k2=0:1:length(temp1)-1
s=s+temp1(k2+1)*exp(-j*(2*pi*f/fs)*k2);
end
TEMP=[TEMP s];
end
FINAL=FINAL+(abs(TEMP).^2);
end
FINAL=1./FINAL;
figure(3)
plot(log(FINAL))
title('Four peaks corresponds to the estimated frequencies')

%ESPRIT(Estimation of signal parameters via rotational invariance techniques)
DATA=Y(1:1:99900);
DATAY=reshape(DATA,9,11100);
DATAY=DATAY(:,2:1:size(DATAY,2));
DATA1=[0 Y(1:1:99900)];
DATA1=DATA1(1:1:99900);
DATAZ=reshape(DATA1,9,11100);
DATAZ=DATAZ(:,2:1:size(DATAZ,2));
RY=cov(DATAY',1);
[E,D]=eig(RY);
sigmaw2=1;
RYZ=cov([DATAY; DATAZ]',1);
RYZ=RYZ(1:1:9,10:1:size(RYZ,2));
CYY=RY-diag(ones(1,size(RY,1))*sigmaw2);
Q=[diag(ones(1,size(CYY,2))); zeros(1,size(CYY,2))];
Q=[zeros(size(Q,1),1) Q];
Q=Q(1:1:9,1:1:9);
CYZ=RYZ-Q*sigmaw2;
[E,D]=eig(CYY,CYZ);
[P,Q]=sort(abs(abs(diag(D))-1));
D1=diag(D);
freqestimation=angle(D1(Q(1:2:8)));
freqestimation=mod(freqestimation,pi)
figure(4)
bar([sort(unwrap(W));sort(unwrap(freqestimation))']')
title('Actual and the Estimated frequencies using ESPRIT algorithm')
```

Chapter 6
Selected Applications in Multidisciplinary Domain

6.1 Multipath Transmission in Wireless Communication

Let the signal $e^{j2\pi ft}$ be transmitted through the multi-path channel. The received signal corresponding to the jth path is obtained as $\beta_j(t)e^{j2\pi f(t-\tau_j)}$. The channel is assumed as the linear and hence the output (received signal) corresponding to the individual paths are combined to obtain the final output as given below.

$$y(t) = \sum_{j=1}^{j=J} \beta_j(t)e^{j2\pi f(t-\tau_j(t))}$$

The response of the time-varying system described by the transfer function $H(f, t)$ to the eigensignal $e^{j2\pi ft}$ is given as $y(t) = H(f, t)e^{j2\pi ft}$. Hence $H(f, t)$ is identified as the following:

$$H(f, t) = \sum_{j=1}^{j=J} \beta_j(t)e^{-j2\pi f\tau_j(t)}$$

$$\Rightarrow h(\tau, t) = \sum_{j=1}^{j=J} \beta_j(t)\delta(\tau - \tau_j(t))$$

The $h(\tau, t)$ is reconstructed using the samples of $h(\tau, t)$ as described below. Let the signal $x(t)$ be sampled using the sampling frequency F_s to obtain the sequence $x(nT_s)$. The signal $x(t)$ is reconstructed back using the discrete samples as described below.

$$x(t) = \sum_{k=-\infty}^{k=\infty} x(kT_s)sinc(\frac{t}{T_s} - k)$$

E.S. Gopi, *Multi-Disciplinary Digital Signal Processing*,
DOI 10.1007/978-3-319-57430-1_6

The response of the time-varying channel to the input signal $x(t)$ is obtained as

$$y(t) = \int_{-\infty}^{\infty} h(\tau, t)x(t - \tau)$$

$$y(t) = \int_{-\infty}^{\infty} \sum_{j=1}^{j=J} \beta_j(t)\delta(\tau - \tau_j(t)) \sum_{k=-\infty}^{k=\infty} x(kT_s)sinc(\frac{t - \tau}{T_s} - k)$$

$$y(t) = \sum_{j=1}^{j=J} \beta_j(t) \sum_{k=-\infty}^{k=\infty} x(kT_s)sinc(\frac{t - \tau_j(t)}{T_s} - k)$$

Let the received signal $y(t)$ be sampled to obtain the discrete sequence $y(nT_s)$ that is obtained as follows:

$$y(nT_s) = \sum_{j=1}^{j=J} \sum_{k=-\infty}^{k=\infty} \beta_j(nT_s)x(kT_s)sinc(\frac{nT_s - \tau_j(nT_s)}{T_s} - k)$$

$$y(nT_s) = \sum_{k=-\infty}^{k=\infty} x(kT_s) \sum_{j=1}^{j=J} \beta_j(nT_s)sinc(n - k - \frac{\tau_j(nT_s)}{T_s})$$

Let $m = n - k$, we get the following.

$$y(nT_s) = \sum_{m=-\infty}^{m=\infty} x((n - m)T_s) \sum_{j=1}^{j=J} \beta_j(nT_s)sinc(m - \frac{\tau_j(nT_s)}{T_s})$$

$$y(n) = \sum_{m=-\infty}^{m=\infty} x_{n-m}h_{n,m}$$

Thus $h_{n,m}$ is the linear time-varying discrete wireless channel and is given as the following.

$$h_{m,n} = \sum_{j=1}^{j=J} \beta_j(nT_s)sinc(m - \frac{\tau_j(nT_s)}{T_s}) \tag{6.1}$$

```
%timevaryingfiltercoef.m
%Let the number of taps be 10;
%Let the sampling frequency be 4000000 Hz.
for n=0:1:1000
for m=1:1:11
    for j=1:1:100
        h(m,n+1)=((rand*2-1)/2)*sinc(m-((rand*2-1)/2)/4000);
    end
end
end
figure
```

```
for i=1:1:11
subplot(4,3,i)
hist(h(i,:))
end
```

The snapshots of the time-varying discrete channel at various time instants n and the corresponding histogram plots are illustrated in Figs. 6.1 and 6.2 respectively.

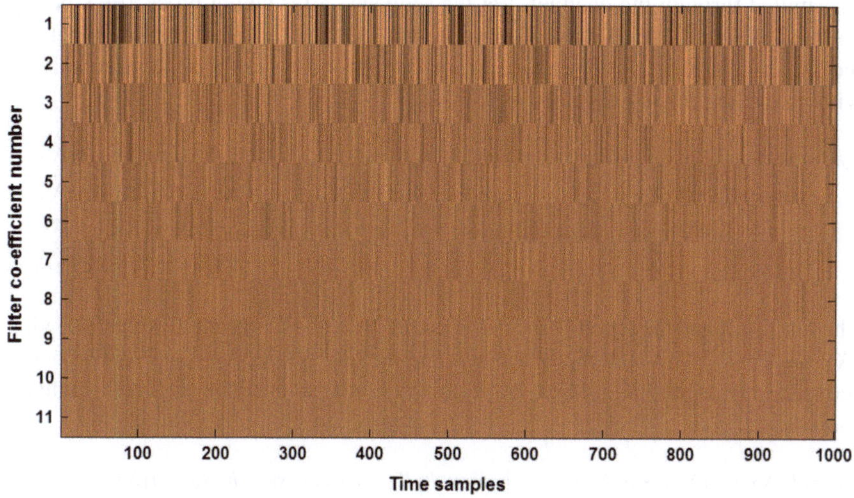

Fig. 6.1 Typical time-varying filter coefficients

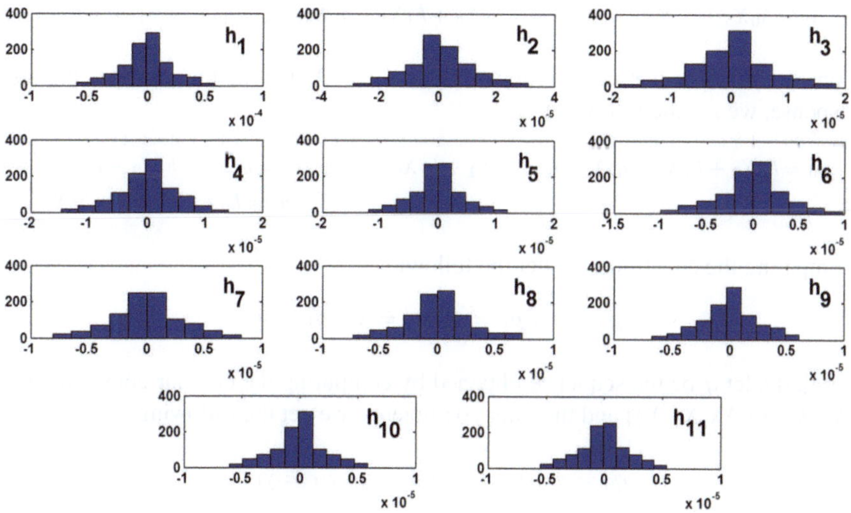

Fig. 6.2 Histogram plot of 11 filter coefficients of the time-varying filter (refer Fig. 6.1)

6.2 Cyclic Prefix in Orthogonal Frequency Division Multiplexing (OFDM)

Let the sequence to be transmitted through the channel be represented as $[x_0\ x_1\ x_2\ x_3\ \cdots x_6\ x_7]$ and the impulse response of the channel be represented as $h_0,\ h_1,\ h_2$. In the case of OFDM, we divide the input sequence into blocks (let us consider the block of four samples) and IFFT of the individual blocks are taken and are transmitted through the channel. Let the sequence after taking block IFFT be represented as $[X_0\ X_1\ X_2\ X_3\ \cdots X_6\ X_7]$. In this $[X_0\ X_1\ X_2\ X_3]$ is the first block and the $[X_4\ X_5\ X_6\ X_7]$. The output sequence corresponding to the block transmission without cyclic prefix is obtained as follows.

$$y_0 = h_0X_0 + h_1X_{-1} + h_2X_{-2},\ y_1 = h_0X_1 + h_1X_0 + h_2X_{-1},\ y_2 = h_0X_2 + h_1X_1 + h_2X_0,$$
$$y_3 = h_0X_3 + h_1X_2 + h_2X_1,\ y_4 = h_0X_4 + h_1X_3 + h_2X_2,\ y_5 = h_0X_5 + h_1X_4 + h_2X_3,$$
$$y_6 = h_0X_6 + h_1X_5 + h_2X_4,\ y_7 = h_0X_7 + h_1X_6 + h_2X_5,\ y_8 = h_0X_7 + h_1X_6 + h_2X_5,$$
$$y_9 = h_0X_8 + h_1X_7 + h_2X_6$$

The first block and the second block after cyclic prefix be represented as the following. $[X_2\ X_3\ X_0\ X_1\ X_2\ X_3]$ is the first block and the $[X_6\ X_7\ X_4\ X_5\ X_6\ X_7]$. The output sequence corresponding to the block transmission after cyclic prefix are obtained as follows.

$$y_0 = h_0X_2 + h_1X_{-1} + h_2X_{-2},\ y_1 = h_0X_3 + h_1X_2 + h_2X_{-1},\ y_2 = h_0X_0 + h_1X_3 + h_2X_2,$$
$$y_3 = h_0X_1 + h_1X_0 + h_2X_3,\ y_4 = h_0X_2 + h_1X_1 + h_2X_0,\ y_5 = h_0X_3 + h_1X_2 + h_2X_1,$$
$$y_6 = h_0X_6 + h_1X_3 + h_2X_2,\ y_7 = h_0X_7 + h_1X_6 + h_2X_3,\ y_8 = h_0X_4 + h_1X_7 + h_2X_6,$$
$$y_9 = h_0X_5 + h_1X_4 + h_2X_3,\ y_{10} = h_0X_6 + h_1X_5 + h_2X_4,\ y_{11} = h_0X_7 + h_1X_6 + h_2X_5,$$

Consider the circular convolution of $[X_0\ X_1\ X_2\ X_3]$ with the channel impulse response, we get the following.

$$p_0 = h_0X_0 + h_1X_3 + h_2X_2,\ p_1 = h_0X_1 + h_1X_0 + h_2X_3,\ p_2 = h_0X_2 + h_1X_1 + h_2X_0,$$
$$p_3 = h_0X_3 + h_1X_2 + h_2X_1$$

Comparing the equations, we get the following.

$$p_0 = y_2,\ p_1 = y_3,\ p_2 = y_4,\ p_3 = y_5$$

Similarly, let q be the sequence obtained by computing the circular convolution of $[X_6\ X_7\ X_4\ X_5\ X_6\ X_7]$ and the filter coefficients, we get the following.

$$q_0 = y_8,\ q_1 = y_9,\ q_2 = y_{10},\ q_3 = y_{11}$$

Actual transmission through the channel involves linear convolution of the sequence with the filter co-efficients. But we observe that when transmission is done with cyclic prefix, the sequence p and q are obtained as the circular convolution of the

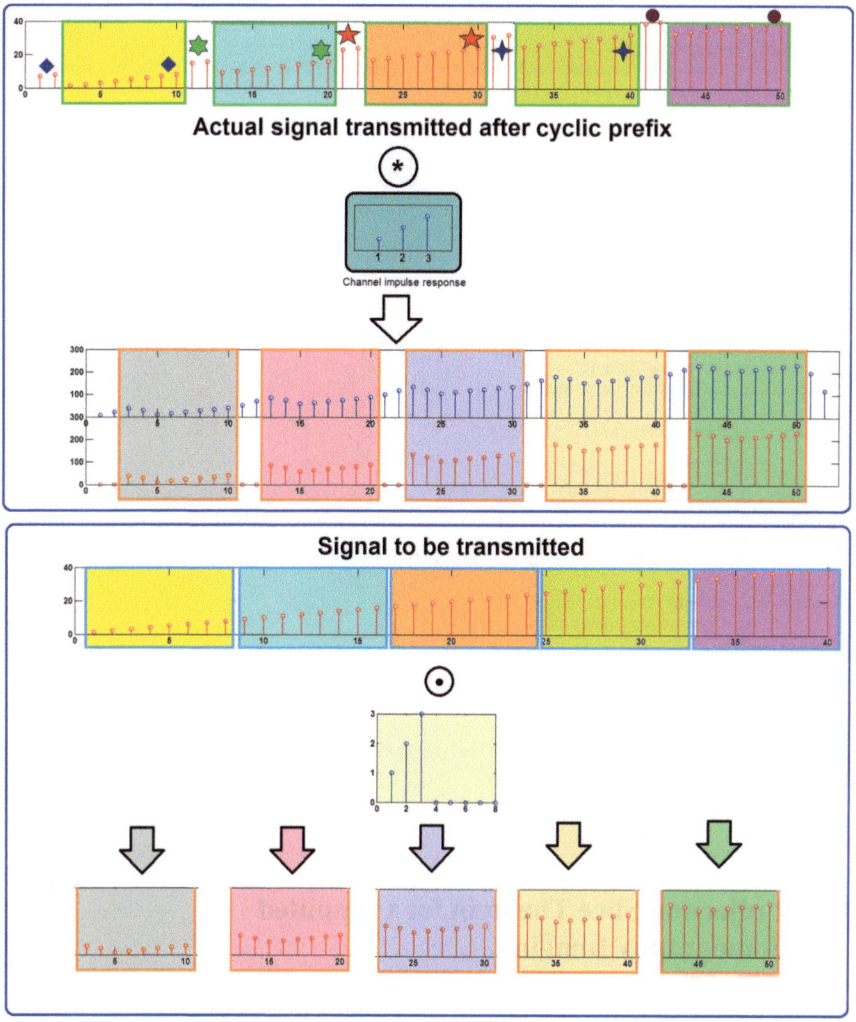

Fig. 6.3 Demonstration of importance of cyclic prefix in OFDM

individual blocks with the filter coefficient. Let the DFT of the filter coefficients h and p be represented as the sequence $H$and $P = x_{block1}H$ respectively. Hence we obtain the transmitted sequence corresponding to block 1 as $\frac{P}{H} = \frac{x_{block1}H}{H} = x_{block1}$. Similarly the transmitted sequence corresponding to block 2 is obtained as $\frac{P}{H} = \frac{x_{block2}H}{H}) = x_{block2}$. We simply take DFT of the received sequence and divide the individual samples with the corresponding DFT of the filter coefficients (after zero padding).

This is equivalently saying that the flat fading is achieved for the individual block (Fig. 6.3).

```
%cyclicprefix.m
%Cyclic prefix in OFDM
%Number of blocks=It is seen outputpart1 and outputpart2 are identical. This demonstrates
%the need for cyclic prefix in OFDM
for i=1:1:5
xblock{i}=(i-1)*8+1:1:(i-1)*8+8;
xblockcp{i}=[xblock{i}(7) xblock{i}(8) xblock{i}];
end
h=[1 2 3];
%Linear convolution with xblockcp
output1=conv([cell2mat(xblockcp)],h);
output1=output1(1:1:50);
output1=reshape(output1,10,5);
outputbefore=reshape(output1,1,size(output1,1)*size(output1,2))
%After ignoring first two columns
output1=output1(3:1:size(output1,1),:);
outputpart1=reshape(output1,1,40)
%Circular convolution with xblock
for i=1:1:5
temp{i}=ifft(fft(xblock{i}).*fft(h,8))
end
outputpart2=cell2mat(temp)
figure
subplot(5,1,1)
stem(cell2mat(xblock))
title('input sequence without cyclic prefix')
subplot(5,1,2)
stem(cell2mat(xblockcp))
title('input sequence after cyclic prefix')
subplot(5,1,3)
stem(outputbefore)
title('Received sequence without removing the first two samples in each block')
subplot(5,1,4)
stem(outputpart1)
title('Received sequence after removing the first two samples in each block')
subplot(5,1,5)
stem(outputpart2)
title('Sequence obtained using circular convolution')
```

6.3 Projection Slice Theorem for Computed Tomography Imaging

It is the method of obtaining the cross section of the object with the array of sources of X-ray and the array of detectors kept at the other side of the object. It is equivalent to obtaining the integration along the lines across the objects to obtain 1D signals. This is repeated by rotating the object by an angle θ and line integration is done to obtain to 1D signal. This is repeated by rotating the object by an angle θ that varies from 0 to 360°. Thus number of 1D signals obtained are $\frac{360°}{theta}$. This is known as Radon transformation. The method of getting the cross section using these 1D signals are known as Inverse Radon Transformation.

Let the cross-section be assumed as the 2D continuous image data represented as $f(x, y)$ (with origin in the middle). Set of parallel lines drawn on the image $f(x, y)$ inclined with an angle θ are represented as $xcos(\theta) + ysin(\theta) = k$, where k is the constant. Let $\theta = 0$, then the set of parallel lines are represented as $x + y = k$ for all k ranges from $-\infty$ to ∞. These are the set of vertical lines drawn on the image $f(x, y)$.

The line integration computed along the lines of these images are represented as the following.

$$r(\theta, l) = \int_{-\infty}^{\infty} \int_{\infty}^{\infty} f(x, y)\delta(xcos(\theta) + ysin(\theta) - l)dxdy$$

The method of obtaining $f(x, y)$ from $r(\theta, l)$ is the inverse Radon transformation as described below. Taking Fourier transformation of the 2D image $f(x, y)$ is computed as the following.

$$F(U, V) = \int_{-\infty}^{\infty} \int_{-\infty}^{\infty} f(x, y)e^{-j2\pi xU} e^{-j2\pi xV} dxdy$$

Let $r(l)$ be the function of l for the fixed θ. The Fourier transformation of the function $r(l)$ is computed as $R(L)$ as follows.

$$R(L) = \int_{-\infty}^{\infty} r(l)e^{-j2\pi Ll} dl$$

$$R(L) = \int_{-\infty}^{\infty} \int_{-\infty}^{\infty} \int_{\infty}^{\infty} f(x, y)\delta(xcos(\theta) + ysin(\theta) - l)dxdye^{-j2\pi Ll} dl$$

$$R(L) = \int_{-\infty}^{\infty} \int_{-\infty}^{\infty} f(x, y)e^{-j2\pi L(xcos(\theta)+ysin(\theta))} dxdy$$

Let $U = Lcos(\theta)$ and $V = Lsin(\theta)$, we get the following.

$$R(L) = \int_{-\infty}^{\infty} \int_{-\infty}^{\infty} f(x, y)e^{-j2\pi Ux+Vy} dxdy$$
$$= F(U, V)$$

$R(L)$ is the 1D function of L obtained from the 2D Fourier transformation of $f(x, y)$ with $U = Lcos(\theta)$ and $V = Lsin(\theta)$. Let us assume $\theta = 0$, we get $U = L$ and $V = 0$. This the vertical line passing through zero in the 2D Fourier transformation. This corresponds to computing $R(L)$ as the following. The value at $R(L)$ is equivalent to $F(U, V)$ at $U = L$ and $V = 0$ with $\theta = 0$. Thus the Fourier transformation of the random transformation r(l) is obtained as the functional values on the 2D Fourier transformation of $f(x, y)$ along the line $U = Lcos(\theta)$, $V = Lsin(\theta)$, which is equivalent to the line described by $UCos(\theta) + Vsin(\theta) - L$. This is known as Projection-slice theorem (refer Fig. 6.4). Thus $f(x, y)$ is computed as follows.

$$f(x, y) = \int_{-\infty}^{\infty} \int_{\infty}^{\infty} F(U, V)e^{j2\pi Ux} e^{j2\pi Vy} dUdV$$

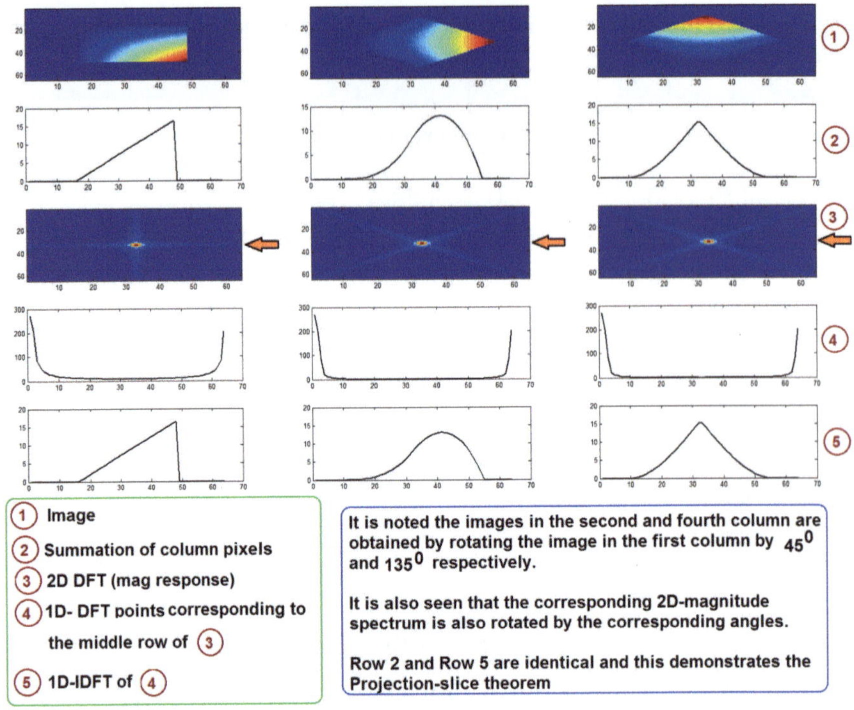

Fig. 6.4 Illustration of projection-slice theorem

Let $U = Lcos(\theta)$ and $V = Lsin(\theta)$, we get

$$f(x, y) = \int_0^{2\pi} \int_0^{\infty} F(Lcos(\theta), Lsin(\theta))|L|e^{j2\pi Lcos(\theta)x} e^{j2\pi Lsin(\theta)y} dLd\theta$$

and hence

$$f(x, y) = \int_0^{\pi} \int_0^{\infty} F(Lcos(\theta), Lsin(\theta))|L|e^{j2\pi Lcos(\theta)x} e^{j2\pi Lsin(\theta)y} dLd\theta +$$

$$\int_{\pi}^{2\pi} \int_0^{\infty} F(Lcos(\theta), Lsin(\theta))|L|e^{j2\pi Lcos(\theta)x} e^{j2\pi Lsin(\theta)y} dLd\theta \quad (6.2)$$

Let $\phi = \theta - \pi$, we get the following.

$$-\int_0^{-\pi}\int_0^{\infty} F(Lcos(\phi+\pi), Lsin(\phi+\pi))|L|e^{j2\pi Lcos(\phi+\pi)x}e^{j2\pi Lsin(\phi+\pi)y}dLd\phi$$

$$=\int_0^{-\pi}\int_0^{\infty} F(Lcos(\phi), -Lsin(\phi))|L|e^{-j2\pi Lcos(\phi)x}e^{-j2\pi Lsin(\phi)y}dLd\phi$$

$$=\int_0^{\pi}\int_0^{\infty} F(Lcos(\theta), Lsin(\theta))|L|e^{j2\pi Lcos(\phi)x}e^{j2\pi Lsin(\phi)y}dLd\phi$$

$$=\frac{1}{2}\int_0^{\pi}\int_{-\infty}^{\infty} F(Lcos(\theta), Lsin(\theta))|L|e^{j2\pi Lcos(\phi)x}e^{j2\pi Lsin(\phi)y}dLd\phi$$

Combining first and second term of (6.2), we get the following.

$$\int_0^{\pi}\int_{-\infty}^{\infty} F(Lcos(\theta), Lsin(\theta))|L|e^{j2\pi Lcos(\theta)x}e^{j2\pi Lsin(\theta)y}dLd\theta$$

For the fixed θ, $\int_{-\infty}^{\infty} F(Lcos(\theta), Lsin(\theta))|L|e^{j2\pi Lcos(\theta)x}e^{j2\pi Lsin(\theta)y}dL$ is the inverse radon transformation and is given as $r(l,\theta) * IFT(|L|)$ and hence

$$f(x, y) = \int_0^{\pi} IFT(R(L,\theta) * |L|dL \tag{6.3}$$

6.4 Warped Discrete Fourier Transformation for Speech Signal

DFT-based frequency components usually suffer with the leakage problem. Hence warped DFT is used to view the frequency domain. In this technique, DFT is obtained by sampling the DTFT in the warped frequency scale. The relationship between the warped frequency w_d and the conventional frequency w is given as the following, where α is the warping parameter. Figure demonstrates the better resolution when warped scale is used to compute samples of DTFT (Figs. 6.5 and 6.6).

$$cos(w_d) + jsin(w_d) = \frac{cos(w) - sin(w) + \alpha}{1 - \alpha cos(w) - \alpha sin(w)}) \tag{6.4}$$

```
%WarpedDFT.m
load speech
%Warping parameter
wp=0.6;
x=SPEECH(5000:6023);
%w2 is the warped frequency
RES1=[];
RES2=[];
W1=[];
W2=[];
for w=-pi:0.01:0
```

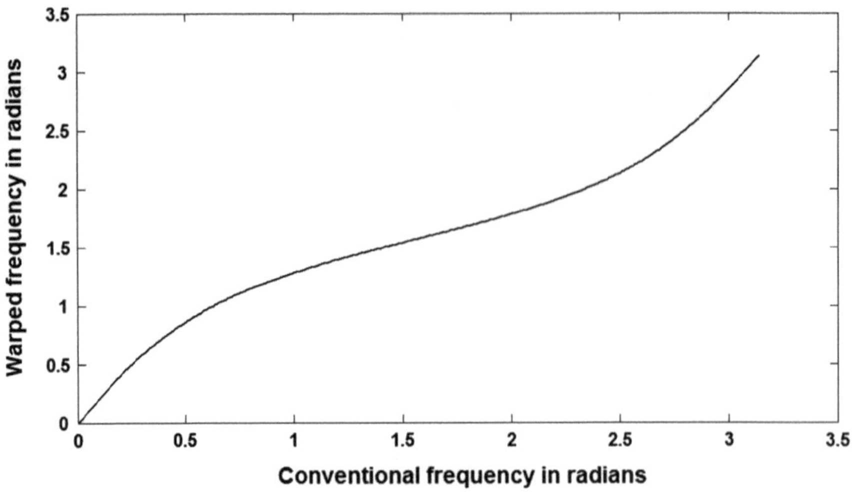

Fig. 6.5 Relationship between the conventional frequency and the warped frequency

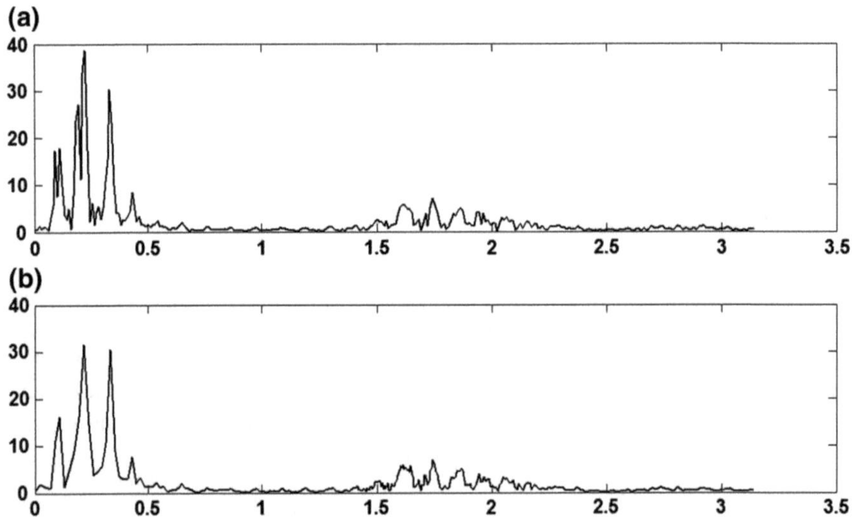

Fig. 6.6 **a** DTFT computed using the conventional scale. **b** DTFT computed using the warped scale

```
w1=angle(cos(w)-j*sin(w));
w2=angle((cos(w)-j*sin(w)+wp)/(1-wp*(cos(w)-j*sin(w))));
s1=0;
s2=0;
for n=0:1:1023
s1=s1+x(n+1)*exp(-j*w1*n);
s2=s2+x(n+1)*exp(-j*w2*n);
end
RES1=[RES1 s1];
RES2=[RES2 s2];
```

```
W1=[W1 w1];
W2=[W2 w2];
end
w=0:0.01:pi;
figure
subplot(2,1,1)
plot(W1,abs(RES1))
subplot(2,1,2)
plot(W2,abs(RES2))
figure
plot(W1,W2)
```

6.5 Nonuniform Discrete Fourier Transformation for the Compressive Sensing Applications

There are applications involved in which the samples are captured in frequency domain (Example:MRI imaging). It is the time-consuming process if all the DFT points are captured in all the frequency points in the frequency. This can be circumvented by capturing the data only for few number of DFT points. This is known as compressive sensing. Given M samples in the N-point DFT, N-point IDFT needs to be computed. This is obtained by interpolating N-point DFT using the available M-point DFT and IDFT is computed to obtain the samples in the time domain. Let the rth sample of N-point DFT sequence of the sequence x be computed directly as $\sum_{k=0}^{k=N-1} x(k)e^{\frac{-j2\pi rk}{N}}$. The values for r are randomly chosen between 0 to $N-1$ (say $r_1\ r_2 \cdots r_M$) and the number of samples are chosen as $M < N$. The continuous spectrum $X(w)$ computed using the M samples are computed using Lagrange interpolation technique is as follows:

1. Obtain $wr_k = \frac{2\pi r_k}{N}$
2. Compute $X(w) = \sum_{p=1}^{p=r} \pi_{q=1,q\neq p}^{q=r} \frac{(w-wr_q)}{w_p-wr_q} X(wr_p)$
3. Sample $X(w)$ to obtain the uniform sampling in frequency domain (DFT) with equally spaced frequencies $w_k = \frac{2\pi k}{N}$
4. IDFT of the samples are computed to obtain the time sequence.

6.6 Nonuniform Sampling of Real Data in Multidisciplinary Applications-Linear Model for Regression

Consider the noisy observation of samples be represented as $t_k^{\prime}s$ corresponding to the input tuning parameter x_k. We would like to estimate the value for the random variable t corresponding to the outcome of the random variable x. The estimate based on MMAE, MMSE, and MAP are obtained as the conditional median, conditional mean, and condition mode. The conditional mean estimate based on MMSE is represented as the following.

$$\hat{t} = E(t/X = x)$$

This can also be solved using the parametric method. In this technique, t is modelled as follows.

$$t = w_0\phi_1(x) + w_1\phi_2(x) + \cdots + w_{M-1}\phi_{M-1} + \varepsilon$$
$$t = W^T\Phi(x) + \varepsilon$$

In this, ε is the random variable associated with noise. It is usually assumed as the Gaussian distributed with mean zero and constant variance v. $w_0, w_1, \cdots w_{M-1}$ are the unknown weight coefficients to be estimated. ϕ is the kernel function. List of commonly used kernel functions are listed below.

- Gaussian basis: $\phi_j(x) = exp^{-\frac{(x-\mu_j)^2}{2s^2}}$
- Sigmoidal basis: $\frac{1}{1+exp^{-\frac{x-\mu_j}{s}}}$

Thus we have identified that t is the random variable with Gaussian distributed with mean $W^T\Phi(x)$ and variance v. Once we get the best estimate for w, to get the estimate of t as the following.

$$\hat{t} = w_0\phi_1(x) + w_1\phi_2(x) + \cdots + w_{M-1}\phi_{M-1} \tag{6.5}$$

The MMSE estimate for w is given as $E(w/t)$. This needs the posterior density function $f(w/t)$, which is not usually available. Hence we estimate w by maximizing the posterior density function with prior density function assumed as uniform distributed. The posterior density function is represented as follows:

$$f(w/t) = \frac{f(t/w))f(w)}{f(t)}$$

6.6.1 Maximum Likelihood Estimation

In this $f(w)$ is the prior density function (which is assumed as uniform distributed) and $f(t)$ is the density function associated with the random variable t (which does not contribute to maximize the posterior density function). Hence w is estimated by maximizing the likelihood function $f(t/w)$. This estimation is known as maximum likelihood estimation. Given the experimental exited values x_k and the corresponding observation t_k, the weight vector W is obtained using the likelihood estimation as follows. Each observation is viewed as the noisy observation and hence each observations are treated as the random variable with mean $W^T\Phi(x_k)$ and variance v. As the observations are independent, we get the joint density function of the random variables $t_1, t_2, \cdots t_N$ (N is the number of observations) as the product of the individual density functions. It is noted that the joint density function is also Gaussian with mean zero and covariance matrix vI, where I is the identity matrix. The optimal w that maximizes this likelihood function is obtained by differentiating the likelihood function with respect to w and equate to zero. Hence we obtain w as the following:

1. Construct the matrix $\mathbf{M} = \begin{matrix} \phi_1(x_1) & \phi_2(x_1) & \cdots & \phi_{M-1}(x_1) \\ \phi_1(x_2) & \phi_2(x_2) & \cdots & \phi_{M-1}(x_2) \\ \cdots & \cdots & \cdots\cdots & \\ \phi_1(x_N) & \phi_2(x_N) & \cdots & \phi_{M-1}(x_N) \end{matrix}$
2. Construct the vector $\mathbf{t} = [t_1 t_2 \cdots t_N]^T]$.
3. The maximum likelihood estimate of column vector W is obtained $W = (M^T M)^{-1} M^T \mathbf{t}$.

6.6.2 Least Square Estimation

We can construct the objective function to estimate W as the one that minimizes $||\mathbf{M}W - \mathbf{t}||^2$. This is known as least square technique. The solution to the least square estimation is identical with that of the maximum likelihood estimate provided that the noise ε is Gaussian. In other words, least square estimation is the maximum likelihood estimate when noise is assumed as Gaussian.

6.6.2.1 $MMSEerror = Bias^2 + Var + Noise$

The estimate value of W depends upon the observation set. To see its consistent, we estimate the vector W using different observation set. Let us assume W_k is the weight vector estimated using the observation set D_k and the corresponding regression equation is obtained as $W_k^T \Phi(X)$. We compute mean of all the curves (obtained using the possible complete range for x); it is obtained using d number of data set to obtain the mean regression curve.

1. If the individual curves are closer to each other, the variance associated with the regression obtained using various data sets is said to be minimum. It is computed for all the possible complete range for x. Let it be Var.
2. MMSE estimate of t, i.e., $E(t/x)$ gives the curve when computed for the possible complete range for x.
3. To see how far the mean curve obtained is closer to the one what we get using MMSE. The mean squared error between the mean curve and the curve obtained using MMSE is the $Bias^2$.
4. It can be that MMSE error obtained using MMSE estimate, i.e. $E(t/x) = Bias^2 + Var + Noise$.

6.6.2.2 Regularization Constant

When we have large number of observations, we divide them into many datasets and the weight vector W is estimated corresponding the individual dataset and the mean regression curve is declared as the final equation. If we have limited dataset, we use regularization factor to avoid over training. This is explained as follows.

It has been seen that maximum likelihood estimate with Gaussian noise model ends up with the least square problem as $||\mathbf{M}W - \mathbf{t}||^2$. Now the objective function is redefined with the inclusion of regularization parameter λ as follows. $||\mathbf{M}W - \mathbf{t}||^2 + \lambda \frac{W^T W}{2}$ Differentiating $(\mathbf{M}W - \mathbf{t})^T(\mathbf{M}W - \mathbf{t}) + \lambda \frac{W^T W}{2}$ with respect to W, we get the following.

$$(\mathbf{M}^T M + \lambda I)W = \mathbf{Mt}$$
$$= W = (\mathbf{M}^T M + \lambda I)^{-1}\mathbf{t}$$

It is noted that when $\lambda = 0$, the estimate is the maximum likelihood estimate (Assuming noise as Gaussian) and also least square estimate (Assuming noise as gaussian).

Thus in practice, regularization is done with various λ and we choose λ corresponding to the minimum value of $Bias^2 + Var$ (refer Fig. 6.7).

6.6.3 *Bayes Estimation*

We estimate W by maximizing the likelihood function assuming that the prior density function is uniformly distributed. Suppose if we choose W to the Gaussian distributed with mean zero (required to get sparse W) and covariance matrix uI, where I is the identity matrix and u is the constant. The posterior density function $f(w/t) = \frac{f(t/w)f(w)}{f(t)}$ is also Gaussian with mean M_P covariance matrix $\mathbf{S_P}$ given as the following.

$$M_P = \frac{1}{v}\mathbf{S_P}\mathbf{M}^T\mathbf{t} \tag{6.6}$$

$$\mathbf{S_P^{-1}} = \frac{1}{u} + \frac{1}{v}M^T \tag{6.7}$$

As $f(w/t)$ is Gaussian distributed, MAP (maximizing a posterior density function) estimate of W is obtained as the mean vector of $f(w/t)$, i.e., M_P, which is given as $(\mathbf{M}^T\mathbf{M} + \frac{v}{u}I)^{-1}\mathbf{M}^T\mathbf{t}$. Thus the Bayes estimate is the regularized estimate with regularization constant given as $\frac{v}{u}$, where v is the variance of the noise and u is the variance of the circularly symmetric prior density function of w, i.e., $f(w)$.

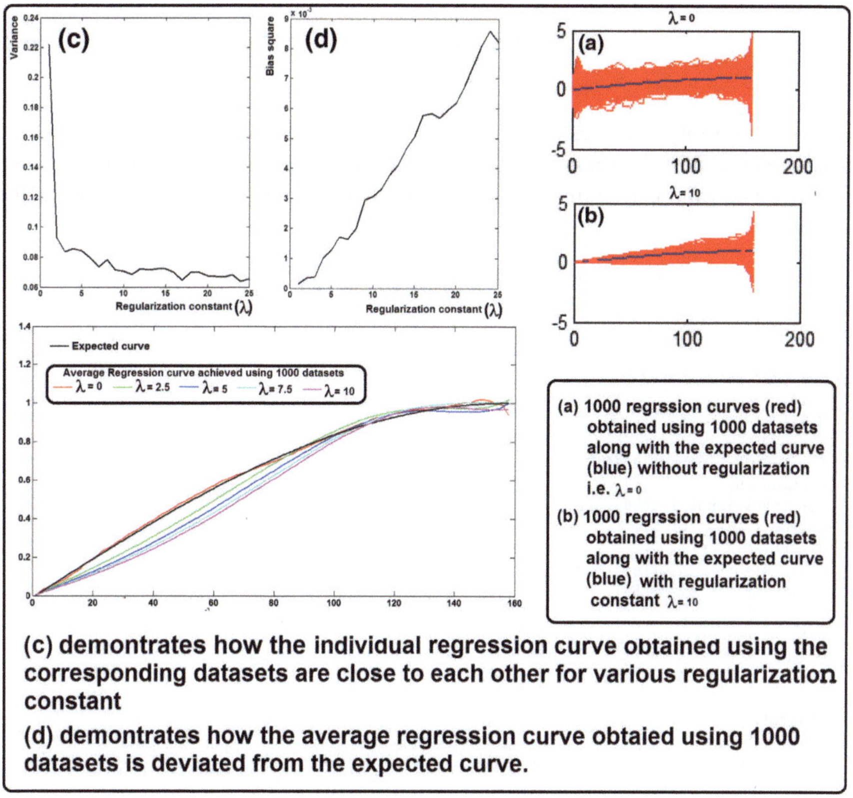

Fig. 6.7 Demonstration of importance of the regularization factor

6.6.4 Kernel Smoothing Function for Regression

Using the Bayes technique, we obtain the estimate for W as given as the conditional mean vector $\frac{1}{v}S_P M^T \mathbf{t}$. Thus the estimate of t given the arbitrary x is given as the following. (refer (6.4) for $\mathbf{M_P}$ and Sect. 6.6.1 for $\mathbf{M}$).

$$\hat{t} = (\frac{1}{v}\mathbf{S_P}\mathbf{M}^T\mathbf{t})^T \Phi(x)$$
$$= \frac{1}{v}(\Phi(x))^T(\mathbf{S_P}\mathbf{M}^T\mathbf{t})$$
$$= \sum_{k=1}^{k=N} \frac{1}{v}\Phi(x)^T \mathbf{S_P}\Phi(x_k)t_k$$

Let $k(x, x_k) = \frac{1}{v}\Phi(x)^T S_P \Phi(x_k)$, we get the following.

$$\hat{t} = \sum_{k=1}^{k=N} k(x, x_k) t_k \tag{6.8}$$

Given the arbitrary x, the estimate of t is obtained as the linear combinations of the observation scalars t_k (training set) and the corresponding kernel functional value $k(x, x_k)$. The valid kernel function $k(x_1, x_2)$ is the one that can be represented as the following:

$$k(x_1, x_2) = \frac{1}{v} \Phi(x_1)^T S_P \Phi(x_2) \tag{6.9}$$

```
%Biasvar.m
k=1;
VAR=[];
BIASSQUARE=[];
for lambda=linspace(0,10,25)
yestmean=0;
for iteration=1:1:1000
x=rand(1,158)*(pi/2);
x=sort(x);
y=sin(x)+2*randn(1,length(x));
x1=linspace(0,pi/2,158);
z=sin(x1);
x=x';
M=[x x.^2 x.^3 x.^4 x.^5 x.^6 x.^7 x.^8 x.^9];
weights=pinv(lambda*diag(ones(1,size(M,2)))+ M'*M)*M'*y';
yest=M*weights;
figure(1)
subplot(5,5,k)
plot(z,'b')
hold on
plot(yest,'r')
YEST{iteration}=yest;
yestmean=yestmean+yest;
end
yestmean=yestmean/1000;
variance=0;
biassquare=0;
for iteration=1:1:1000
variance=variance+(YEST{iteration}-yestmean).^2;
end
variance=sum(variance)/(1000*158);
VAR=[VAR variance];
figure(2)
biassquare=(yestmean-z').^2;
plot(yestmean,'r')
hold on
plot(z,'b')
biassquare=sum(biassquare)/158;
BIASSQUARE=[BIASSQUARE  biassquare];
k=k+1;
end
figure
plot(VAR)
title('Variance versus regularization constant')
figure
plot(BIASSQUARE)
title('Bias square versus regularization constant')
```

The list of commonly used kernel functions $k(x_1, x_2)$ (Applicable for vectors) are listed below.

1. Inner product kernel: $x_1^T x_2$
2. Gaussian kernel: $e^{\frac{-||x_1-x)2||^2}{c_1}}$
3. Polynomial kernel: $(x_1^T x_2 + c_2)^{c_3}$
4. Power exponential: $(e^{\frac{-||x_1-x_2||^2}{c_4}})^{c_5}$
5. Hyperbolic tangent: $tan(c_6(x_1^T x_2) + c_7)$
6. Cauchy: $\frac{1}{1+\frac{||x_1-x||^2}{c_8}}$
7. Inverse multiquadric: $\frac{1}{\sqrt{||x_1-x_2||^2+c_9^2}}$

List of m-files

demonstratereconstruction.m Reconstruction of sampled signals.

samplingtheorem.m Nonoverlapping of spectrum in DTFT.

circconvusingDFT.m Circular convolution using DFT.

dit.m Decimation in Time FFT.

executedit.m Comparison of time required to compute DFT using the direct method and using DIT.

dif.m Decimation in Frequency FFT.

executedif.m Comparison of time required to compute DFT using the direct method and using DIT.

subband.m Sub-band filters.

responsetoperiodicsequence.m Response of the discrete system to the periodic sequence.

complexgeometry.m Demonstration of the geometrical interpretation of computation of magnitude and phase response of the discrete transfer function.

magrespzplot.m Called by the m-file complexgeometry.m.

plotbuttermag.m Plot the magnitude response of the Butterworth analog filter.

plotchebymag Plot the magnitude response of the Chebyshev analog filter.

CN.m Called by plotchbymag.m.

butterworthorder.m Compute the order of the Butterworth filter.

digitalbutterworth.m Converts analog Butterworth filter to digital filter.

impulses2z.m Impulse invariant mapping.

bilinears2z.m Bilinear transformation mapping.

E.S. Gopi, *Multi-Disciplinary Digital Signal Processing*,
DOI 10.1007/978-3-319-57430-1

digitalchebyshev.m Converts analog Chebyshev filter to digital filter.

chebyshevorder.m Compute the order of the Chebyshev filter.

ButterworthHPFdemo.m Demonstration of digital Butterworth high-pass filter.

digitalbutterworthHPF.m Function called by ButterworthHPFdemo.m.

chebyshevHPFdemo.m Demonstration of digital Chebyshev high-pass filter.

digitalchebyshevHPF.m Function called by chebyshevHPFdemo.m.

IIRBPFDEMO.m Demonstration of digital bandpass filter.

digitalBPF.m Function called by IIRBPFDEMO.m.

IIRBRFdemo.m Demonstration of digital bandpass filter.

digitalBRF.m Function called by IIRBRFDEMO.m.

realizeiir.m Realization of IIR filter using Directform-1 and Directform-II.

FIRdemo.m FIR filter demonstration

firtype1.m Called by FIRdemo.m.

firtype2.m Called by FIRdemo.m.

firtype3.m Called by FIRdemo.m.

firtype4.m Called by FIRdemo.m.

minmaxFIR.m Demonstration of relationship between min-phase, max-phase, and all pass FIR filters.

effectoftruncation.m Demonstration of magnitude response of the truncated filter.

truncimpLPF2FIR1FIR3.m Illustration of obtaining the low-pass filter coefficients for FIR 1 and FIR 3 filter from the truncated impulse response of the ideal low-pass filter.

truncimpLPF2FIR2FIR4.m Illustration of obtaining the low-pass filter coefficients for FIR 2 and FIR 4 filter from the truncated impulse response of the ideal low-pass filter.

HPFtype1.m Illustration of obtaining the Type 1 high-pass filter coefficients.

HPFtype3.m Illustration of obtaining the Type 3 high-pass filter coefficients.

BPFT1LPFT1HPF.m Bandpass filter constructed using Type 1 LPF and Type 1 LPF.

BPFT3LPFT3HPF.m Bandpass filter constructed using Type 3 LPF and Type 3 LPF.

BPFT3LPFT1HPF.m Bandpass filter constructed using Type 3 LPF and Type 1 LPF.

BPFT1LPFT3HPF.m Bandpass filter constructed using Type 1 LPF and Type 3 LPF.

BRFT1LPFT1HPF.m Band-reject filter constructed using Type 1 LPF and Type 1 HPF.

BRFT3LPFT3HPF.m Band-reject filter constructed using Type 3 LPF and Type 3 HPF.

BRFT1LPFT3HPF.m Band-reject filter constructed using Type 1 LPF and Type 3 HPF.

BRFT3LPFT1HPF.m Band-reject filter constructed using Type 3 LPF and Type 1 HPF.

reduceripplesLPF.m Windowing technique to reduce ripples for FIR low-pass filter.

reduceripplesHPF.m Windowing technique to reduce ripples for FIR high-pass filter.

reduceripplesBPF.m Windowing technique to reduce ripples for FIR bandpass filter.

reduceripplesBRF.m Windowing technique to reduce ripples for FIR band-reject filter.

identicalmagresfilters.m FIR filters having identical magnitude responses.

multiratedemo.m Demonstration of decimation and interpolation.

interpolyphase.m Interpolation using polyphase realization.

decimatepolyphase.m Decimation using polyphase realization.

QMFdemo.m Quadrature mirror filter demonstration.

QMF.m Called by QMFdemo.m.

transmultiplexer.m Demonstration of transmultiplexer.

wssrp.m Demonstration of wide sense stationary random process.

forbackcoef.m Computation of forward and backward prediction coefficients.

adapsysiden.m Adaptive filter for system identification.

adapfiltnoise.m Adaptive filter for noise removal.

pisarenko-music-esprit.m Demonstration of Pisarenko, MUSIC, and ESPRIT algorithms for spectral estimation.

timevaryingfiltercoef.m Time-varying filter coefficients of the wireless channel.

cyclicprefix.m Demonstrating importance of cyclic prefix in OFDM.

WarpedDFT.m Usage of warped DFT for signal analysis.

Biasvar.m Demonstration of importance of regularization factor in regression.

Index

E.S. Gopi, *Multi-Disciplinary Digital Signal Processing*,
DOI 10.1007/978-3-319-57430-1